Life's Chemical Secret

The MIT Press's publishing mission benefits from the generosity of our donors, including Mike Dornbrook.

Life's Chemical Secret

The Origin of Purpose and Mind

Addy Pross

The MIT Press
Cambridge, Massachusetts | London, England

The MIT Press
Massachusetts Institute of Technology
77 Massachusetts Avenue
Cambridge, MA 02139
mitpress.mit.edu

The MIT Press would like to thank the anonymous peer reviewers who provided comments on drafts of this book. The generous work of academic experts is essential for establishing the authority and quality of our publications. We acknowledge with gratitude the contributions of these otherwise uncredited readers.

This book was set in ITC Stone Serif Std and ITC Stone Sans Std by New Best-set Typesetters Ltd. Printed and bound in the United States of America.

Library of Congress Cataloging-in-Publication Data

Names: Pross, Addy, 1945– author
Title: Life's chemical secret : the origin of purpose and mind / Addy Pross.
Description: Cambridge, Massachusetts : The MIT Press, [2026] | Includes bibliographical references and index.
Identifiers: LCCN 2026001695 (print) | LCCN 2026001696 (ebook) | ISBN 9780262054836 hardcover | ISBN 9780262055352 pdf | ISBN 9780262055369 epub
Subjects: LCSH: Life—Origin | Chemistry—Philosophy | Mind and body—Philosophy
Classification: LCC QH325 .P88 2026 (print) | LCC QH325 (ebook)
LC record available at https://lccn.loc.gov/2026001695
LC ebook record available at https://lccn.loc.gov/2026001696

10 9 8 7 6 5 4 3 2 1

EU Authorised Representative: Easy Access System Europe, Mustamäe tee 50, 10621 Tallinn, Estonia | Email: gpsr.requests@easproject.com

For Nella, Guy, and Tamar, always

Contents

Preface

Here we are, well into the twenty-first century, yet modern science continues to struggle with a profound contradiction in the natural order. On one hand, the nonliving world—from atoms to galaxies—follows the rules of physics with astonishing precision. This objective, materialist view of nature, which has guided scientific thinking since the days of Galileo and Newton, holds that reality can be measured, modelled, and explained.

But the living world doesn't fit into that materialist mold. Living things express characteristics that are incompatible with a view of nature as purely objective. Purpose, desire, intentionality, mind, consciousness—these are anything but objective.

How could natural processes have led to such unnatural consequences? What is the physical basis for life's seemingly nonphysical characteristics? Biology's response to the question has been awkwardly uncertain, ranging from simply classifying those characteristics as "emergent properties," or deeming them "illusory." Carl Woese, the luminary biologist, remarked recently that modern biology is “no more fully understood in principle than physics was a century ago,” claiming its theoretical foundations are in pressing need of reformulation. And physical scientists, in whose jurisdiction the problem properly lies, have been equally puzzled. Erwin Schrödinger, the renowned twentieth-century physicist, in referring to consciousness, stated it pointedly: “consciousness cannot be accounted for in physical terms . . . it is absolutely fundamental. It cannot be accounted for in terms of anything else.”

The scientific stakes are high. The philosopher Thomas Nagel argued that if the modern physicalist view of life, supported by several decades of detailed molecular biological revelations, cannot explain mind-related features of our world, then the entire naturalistic worldview—encompassing biology, evolutionary theory, and cosmology—may need radical revision. Perhaps the tools science has used to understand the inanimate world aren’t sufficient to explain life.

But the fact that life on Earth almost certainly emerged from inanimate beginnings some 3.5–4 billion years ago means that life and nonlife do connect. Chemistry did become biology. But how? A theory of matter that describes the inanimate world is well established, but an equivalent theory able to explain the living world with life’s unique physical and nonphysical characteristics still eludes us.

This short book seeks to address this fundamental question by describing how one missing piece of physical information may have been largely responsible for the conceptual dead end in which the natural sciences seem to have been stuck for close to a century.

More than a century ago, the physicist Willard Gibbs helped define the familiar thermodynamic states of matter—solid, liquid, and gas. Only recently, however, has chemistry recognized that matter can also exist in a fundamentally different, nonthermodynamic state: a kinetic state. In this regime, molecules are not passively at equilibrium but are in continual, energy-driven motion, even as the system as a whole remains stable, seemingly unchanged. This kind of stability is termed *dynamic kinetic stability* (DKS), admittedly a technical mouthful. To help understand the term, think of a fountain—energy driven, in perpetual motion, yet appearing tantalizingly still.

This DKS state is not a mere theoretical curiosity. A growing number of real-world materials exhibit such kinetic stability, and this discovery has far-reaching implications. Life, this book argues, is just the most striking example of matter in that kinetic dimension. And in a remarkable and unexpected development, the reality of kinetic states may have opened the door to the physical dimension of mind.

Based on the recognition of life as a kinetic state of matter, a physical/chemical theory of life can now be outlined. Reductionist methodology, the mainstay of scientific understanding for centuries, may be applicable to biology's core questions after all:

- Why is life so complex—far more than physics alone would predict?
- How did consciousness arise in a universe built from particles and fields?
- Does evolution have direction, a driving force?

In this book, I attempt to present a relatively compact and largely qualitative description of these ideas, building on my early theoretical suggestion that kinetic states of matter likely exist, a prediction subsequently confirmed experimentally. The physical basis for life's existence appears to be emerging from the metaphysical shadows, with profound implications for our understanding of purpose and mind.

I
The Problem

1
Science's Black Hole

> At the core of all well-founded belief, lies belief that is unfounded.
>
> —Ludwig Wittgenstein

Let me begin by stating the obvious: modern science divides nature into physics, chemistry, and biology. But this division reflects not how nature is, only how we have learned to study it. Science recognizes nature's hierarchical and diverse character, but only up to a point. As understanding deepens, disciplinary boundaries dissolve, and the separate provinces of science begin to reunite. Physics and chemistry have already completed this journey. Nature itself was never divided; only our descriptions were.

Biology's status, however, is conceptually problematic. Though the issue tends to be marginalized, a deep conceptual chasm continues to separate the physical and biological sciences. On one hand, natural scientists all agree that biological entities are comprised of nothing but physical and chemical material. Vitalism, the view that living organisms are governed by a "vital force," was largely vanquished at the turn of the twentieth century. As Ernst Mayr, the Harvard evolutionary biologist put it some years ago, "no biologist alive today would want to be classified as a vitalist."

And yet, there is a striking disconnect between living and nonliving, one that is obvious even to a child: toy dogs and real ones behave very differently. What is the physical basis for life's "aliveness," that intangible quality that somehow emanates from life's intricate network of physical and chemical processes? What special chemistry underpins that "aliveness"—the chemistry that abruptly disappears the moment death displaces life and "regular" chemistry takes over? How can these chemical differences be understood? Physics and chemistry continue to struggle to bridge that conceptual gap. And biologists, finding themselves intellectually abandoned by their physicist colleagues, took the only viable option seemingly available: to treat life as a fundamentally distinct kind of phenomenon. An "autonomy of biology" approach to the discipline thus became established, one that tacitly accepted, and largely set aside, the unresolved tensions within a strictly physicalist account of life.

But such a situation is clearly unsatisfactory. Here we have two major sciences, physics and biology, each anchored in material reality, irrevocably joined at the hip, yet seemingly intrinsically incompatible. One is reminded of the two mismatched roommates in the play *The Odd Couple*—unable to separate, but unable to live together. And the tensions between the disciplines (along with their respective scientists) go beyond Lord Rutherford's famous pejorative that "all science is either physics or stamp collecting," with biology haughtily relegated to the latter category.

There are of course many points of agreement between the two sciences. Even quarrelling couples agree on some things. Both sciences acknowledge that the material world, both living and nonliving, is made up of the same fundamental atomic building blocks. Both sciences acknowledge that atoms can join up in a multitude of different ways to make up an almost infinite array of possible molecules. And both sciences also agree that molecules can undergo self-assembly to generate "stuff." But from that point, dramatic differences become apparent.

Consider self-assembly, a common phenomenon within both the physical and biological worlds. Chemical stuff routinely assembles into lifeless forms, such as crystals or amorphous rocks. But in unique circumstances, it can also self-assemble into living forms, such as bacteria, birds, and bees. The difference between those two kinds of self-assembly is no less than astounding. In the nonliving world, when a solution of, say, regular table salt (sodium chloride) becomes saturated, the moment a microcrystal has formed, invisible to the naked eye, that microcrystal attracts more of the sodium and chloride ions present in the solution, and the microcrystal grows. Each ion locks into place as determined by physical law, leading to the formation of a highly ordered three-dimensional crystal lattice. We will discuss in due course the nature of the physical law responsible for this ordering process, but at this point suffice it to say that it is one of the most important laws of nature—the Second Law of Thermodynamics. A snowflake is another example of that kind of self-assembly. A magnified image of a snowflake shows the formation of exquisite structures displaying remarkable symmetry. Such structures demonstrate that nature can direct processes and does so in an extraordinarily precise fashion. The directing abilities of the Second Law are unambiguous and well understood.

But then there is biological self-assembly as reflected, for example, in the process called ontogeny. Here, a fertilized egg, similar in size to the microcrystal in the crystallization experiment, also grows over time. It attracts material from the environment, just like that microcrystal—but in this case, it grows into an embryo and eventually into an adult, whether a human, a frog, or a butterfly. That transformation is clearly directed, just as snowflake formation is directed, but here the self-assembly is directed elsewhere and occurs in a dazzlingly implausible way. That miniscule single-cell aggregate, through a process of continuing assembly, eventually becomes billions of differentiated cells that make up the organs of the particular organism it represents—hearts, kidneys, eyes, ears, bones, limbs. And depending on the detailed molecular structure of several of the molecules within that primal cellular structure, the ultimate morphological outcome of that assembly process is largely predetermined and can be staggeringly different, resulting in the emergence of a human, a frog, or a butterfly. Broadly speaking then, we see before us two kinds of self-assembly. In both, the beginnings can be found in a microscopic aggregate, but the self-assembly that ensues in the two cases could not be more distinct. They differ in their form, their physics, their phenomenological qualities, their diversity. Most importantly, one aggregate is alive, the other isn't.

What causes self-assembly to follow such distinctly different material paths? Science currently offers no clear answer. Both processes are consistent with the Second Law of Thermodynamics, but only in one case do we have a fundamental understanding of how the Law operates. We can cause water to self-assemble into ice, we can induce crystals of salt to precipitate from a saturated salt solution—all easy processes to carry out, all physically straightforward, all expressions of the Second Law in action. But when it comes to biological self-assembly, we remain in the dark. What nature does routinely, deliberately, and with exquisite precision, we cannot even begin to mimic. We have no idea how to take a solution of life's established building blocks—proteins, nucleic acids, sugars, lipids, some metal ions, and such, obtained in some manner, perhaps by extracting them from dead cells—and get them to re-assemble in the biological way to construct a living thing. Actually, self-assembly of those biomolecules can be induced, but it won't be *biological* self-assembly, as observed during ontogeny; it will be regular *physical* self-assembly. Self-assembly of life's building blocks when carried out in a laboratory will result in the formation of inhomogeneous "dead" clumps of matter, not the exquisitely precise biological self-assembly that leads toward a living creature. The special kind of self-assembly that leads to a living organism only takes place when you start off with the appropriate initial aggregate—a living fertilized egg—and continue to supply it with appropriate raw materials, all the while maintaining suitable conditions that ensure the living entity remains alive. We recognize that two groups of molecules in particular, nucleic acids and proteins, are central to this other kind of self-assembly, but precisely how the process takes place remains a mystery.

So, with the seemingly simple phenomenon of molecular aggregation, we are already faced with a major scientific riddle: Even though biological self-assembly takes place routinely before our very eyes, the chemical principles that govern such self-assembly remain largely unknown and we remain unable to duplicate the process from inanimate beginnings. Life, as a physical phenomenon, is characterized as much by its special organization as by its constituents. How is it that nature is able to self-assemble life's ingredients in that special organized fashion, the hallmark of all living things, yet we scientists, with all our scientific knowledge on matter and material processes, are unable to do so? We are not even sure how to begin. The reason for the difficulty is clear: we don't understand the process. And the central reason we don't understand the process is because we are lacking a physical theory of life.

But the life mystery does not stop there. When the self-assembly associated with living things has come about, it results in properties that undermine the most basic tenets of physical epistemology—first and foremost, agency. All living things appear purposeful, able to act on their own behalf, and do so through their internal organization, which itself is purposeful. Think about that—organization that produces matter with an underlying agenda. How is that possible? Ice crystals or salt crystals that result from physical self-assembly certainly show no signs of purposeful organization. Those structures pack in a way that minimizes the crystal's energy, just as balls rolling downhill seek the point of lowest potential energy. Yet even the simplest living thing, a bacterial cell, is not organized as a structure seeking to lower its energy. In fact, the reverse is true. The cell is in an energized state, continuously tapping into an external source of energy to maintain its purposeful, deliberate organization. The biological cell is effectively the world's most sophisticated and elaborate copying machine. In a cell, we find a physical entity, composed of several million functional molecules and molecular aggregates, that faithfully reproduces itself, with inbuilt proofreading to correct for inadvertent copying errors. All that purposeful organization has a clear and identifiable goal. As the noted French biologist François Jacob put it: *the dream of every cell to become two cells*.

The extraordinary material behavior of living things may also be viewed from the perspective of existing material states of matter. Any decent physics or chemistry textbook will inform you that there are three states of matter: solid, liquid, and gas (not including two more exotic ones—the plasma state, which can be observed at extremely high temperatures, and Bose-Einstein condensates, which can be observed at extremely low temperatures). Living things, however, are decidedly material, yet they are not formally recognized as state of matter, even though they are all around us. How can it be that a material form that has overwhelmed our planet has no definable physical or chemical characterization? Indeed, the leading physicists of the twentieth century—amongst them Niels Bohr, Erwin Schrödinger, and Eugene Wigner—were deeply puzzled by the life phenomenon and were led to conclude that there appear to be laws of physics yet to be discovered to account for their nonmaterial characteristics.

In the late 1970s, Belgian physicist Ilya Prigogine made a significant contribution to the topic by extending thermodynamic principles to *nonequilibrium* systems, namely systems that are not in their lowest energy state. He showed that systems in a nonequilibrium state can self-organize and maintain structures *away* from equilibrium, despite the natural thermodynamic drive toward equilibrium. Familiar everyday examples include whirlpools, hurricanes, and so-called Bénard cells. Whirlpools, hurricanes, and Bénard cells can only exist provided an energy source is present to maintain them. Bénard cells can typically be generated in a layer of liquid that is heated from below. Doing so results in a striking pattern of hexagonal convection cells that can be maintained as long as the heat supply is continued. All such nonequilibrium structures are termed *dissipative structures*, as they continually dissipate energy in order for those nonequilibrium structures to be maintained.

And yet, these ideas alone were unable to resolve the life dilemma. While Prigogine's landmark ideas on nonequilibrium thermodynamics did manage to narrow the conceptual gap separating living and nonliving, they did not lead to the much sought-after breakthrough. The conceptual gap separating living and nonliving systems remained unresolved.

But returning to our states-of-matter discussion, we could classify life as a *fourth* state of matter alongside the three conventional states: solid, liquid, and gas. (This classification might initially seem inappropriate because the three conventional states of matter are *thermodynamic states*. The living state, being *open*—meaning that it undergoes material and energetic exchange with its environment—is *not* a thermodynamic state. But for the purpose of our analysis, its characterization as a state of matter will prove to be a useful one.) These three thermodynamic states are, of course, well understood despite their different properties. Water in its liquid phase is quite different from water in its solid form, ice, and very different again from water in its gas state, steam. In fact, water in the gas phase is invisible. Hard to be more different than that.

But what is important is that the three thermodynamic states, despite being so different, are well understood thanks to a theory of matter, a kinetic theory, formulated in the mid-nineteenth century. The theory was based on the idea that matter is made up of constantly moving particles and that the energy of the particles, their kinetic energy, depended on the temperature of the material sample. Due to attractive forces that exist between individual particles, the observed state of matter depends directly on the level of kinetic energy. At low temperatures, when the energy is low, the attractive forces between the particles win out and the material exists as a solid. At intermediate temperatures, when the particles are more energetic, the attractive forces are unable to keep the particles locked into one static configuration. But those forces do keep the particles in close proximity, resulting in formation of the liquid state. At even higher temperatures, the kinetic energy of the particles is high, enabling them to completely break away from one another, leading to the gas state. Thus, thanks to the kinetic theory of matter, we understand the properties of each of the three states, their relationship to one another, and, most importantly, based on that kinetic theory, we know how to convert any one state into any other state through what are termed phase transitions. Indeed, that ability to interconvert the three states at will is the direct expression of our understanding.

So where does the living state fit into this scheme? Well, it doesn't. Superficially the living state appears solid-like in its structural scaffolding and liquid-like in its internal chemistry. Yet, regardless of the precise state description, the means of interconverting living and nonliving states—that direct expression of material understanding—remains out of reach, a scientific mystery. It is of course all too easy to convert living matter into "dead" stuff. But how does one coax the nonliving state back into the living one—without Dr. Frankenstein's assistance? That remains a mystery. So let's make a start.

As already noted, having possession of the material building blocks that make up the living state is insufficient. So, if we classify the living state as a fourth state of matter, we do not know how to induce a phase transition into that fourth state. Puzzlingly, physics, the science of matter and energy, seems to have given up on answering that central question, one that lies at the very heart of biology. Success on this issue would be an important first step toward enabling physics and biology, that odd couple, to fully connect, as fundamentally they must.

* * * * * *

Given the pragmatic division of the material world into physical and biological groupings, let us now consider how physical laws, formulated for physical systems, and biological laws formulated for biological ones, might relate to one another.

Physics, of course, deals with the material world in its entirety, while biology just deals with the living segment of it. Of course, it goes without saying that the laws of physics apply to biological systems as well. The force of gravity operates on living organisms just as it does on nonliving ones. It could not be otherwise. But when it comes to biological laws, the connection to physical laws is less clear.

There are some laws that apply within the biological world that appear inapplicable to the more general physical world. In some cases, that might be considered acceptable, as biological organization can lead to what are termed *emergent properties*. We know too well that at higher levels of organization, new properties can come about that are not discernible at lower organizational levels. To illustrate, general principles of ecology and physiology have no relevance in the nonliving world. There is an ecological principle, the *competitive exclusion principle,* which states that ecological differentiation is the necessary condition for coexistence of two species occupying the same ecological niche. That's a biological principle with no direct bearing on the physical world and so does not contradict any physical principles. Turning to individual bodies, the concept of homeostasis is a fundamental one in physiology. Scott Turner, a State University of New York biologist, has labelled homeostasis as Biology's Second Law (Biology's First Law being natural selection)—the clear tendency of living things to maintain themselves in a particular physiological state, for example, constant organismic temperature or constant intracellular ion concentration. Homeostasis is not a law of physics, but it does not contradict any law of physics.

While we can feel comfortable with the existence of biological laws that are not applicable to physical systems, we would not expect such laws to be *inconsistent* with established physical laws. In fact, it would come as a surprise if we were to find that key laws of physics and biology were contradictory. But that precisely, is the current reality.

Consider the concept of function, one familiar to all biologists. Indeed, physiology, a sub-discipline of biology, is almost entirely built upon that concept. How does a particular organ or organelle work? No one would deny that the function of kidneys is to filter metabolic waste from the blood stream, or that the function of ribosomes is to make proteins. And proteins, which have been produced by ribosomes, have their functions as well. For example, the role of the two proteins, myosin and actin, is to enable muscle function. The list of functions in biology is effectively endless. Biology without the concept of function would be like physics without the concept of force or energy. Without its concept of function, biology would make little sense.

But if you step back for a moment and think of living things as just particular material forms, the question immediately arises: how can function in the natural world actually exist? Any function? As any physicist will tell you, in the inanimate world there are no functions, just nature's objective laws. Rocks, sand, rivers, atoms, electrons have no function. These inanimate entities simply obey the laws of physics and chemistry in which function has no role. So, we have here a troubling paradox. The biological world cannot be discussed *without* the concept of function, whereas within the physical world the concept of function not only does not exist, but is seemingly inconsistent with the "nature is objective" rule. That rule was at the heart of the modern scientific revolution that changed scientific thinking some four centuries ago.

The above stated inconsistencies between physical and biological law do not end there. Consider the question of *complexity* as it manifests in the living and nonliving worlds. Living things are complex, bewilderingly so, and magnitudes more complex than any existing inanimate entity, even spectacularly large ones. Martin Rees, a former UK Astronomer Royal, pointed out that giant astronomical entities, despite their size, may not be complex, contrary to what we might intuitively think. Stars, for example, are enormous bodies, many billions of times larger than our planet, but structurally speaking, they are relatively simple. Their spectacularly high temperature, as high as several million degrees Celsius, is such that complex molecules cannot exist under those conditions, so stars are fundamentally composed of hydrogen and helium nuclei with smaller amounts of heavier elements sprinkled throughout. Big, but simple.

Life however is anything but simple. The *smallest* living things may be effectively more complex than the *largest* nonliving ones! That's quite extraordinary. How can we rationalize that staggering difference in complexity? Science currently has no definitive answer. And even more remarkably, bacteria, already highly complex, are just the beginning of an evolutionary process toward further increases in complexity. Simon Conway Morris's memorable phrase captures the point well: *once there were bacteria, now there is New York*. Complexity in the two worlds, physical and biological, couldn't be more distinct. Yet neither physicists nor biologists have a clear understanding of the nature of the very special organized kind of complexity that is found in living things. And, more problematically, how it was able to arise.

The conflicting character of physics and biology also becomes glaringly obvious through a comparison of the two grand nineteenth-century laws—the Second Law of Thermodynamics, which applies to matter in general, and Darwin's theory of evolution, which addresses change within living populations. While the Second Law explains why material forms are generally directed *downward,* toward greater stability, toward uniformity and disorder, evolutionary change in living systems seems to be directed *upward,* toward greater complexity, toward increasingly organized *unstable* entities. How could such central laws of the universe be seemingly contrarian?

And it's not just the *direction* of change in the two worlds that differs; it is the very nature of the change that differs. The Second Law predicts that the universe is destined to grind toward its ultimate demise, toward what physicists ominously term "heat death." But through the existence of that other world, matter became alive—literally—at least it did so on at least one location in the universe. But how was such a transition possible? What caused the universe to awaken? How could function and purpose have emerged from an unfolding universe headed irrevocably toward heat death? Is it any wonder that physics and biology remain awkward bedfellows? The two landmark laws that revolutionized our understanding of nature, one with regard to the inanimate world, the other with regard to the living world, do not cohabit well. Until science is able to come up with a physical/chemical theory of life that can bridge these two conceptually distinct worlds, biology will remain a scientific island, one conceptually disconnected from physics and chemistry, one without a sound theoretical foundation. Until that theory of life is formulated, living forms will remain the inexplicable state of matter.

The unavoidable conclusion: biology's theoretical base cannot be sought within biology but needs to be sought within physics and chemistry. In fact, somewhat paradoxically, the life question is not in essence a biological problem, but rather a physical/chemical one. It will no longer suffice for biology to formally recognize life's material basis and then proceed to put to one side the necessary consequences of acknowledging that material basis. Biological theory will need to conform to physical and chemical law in all respects, with contradictions between established biological and physical/chemical laws addressed and eliminated. Theoretical biologist Stuart Kauffman recently claimed the existence of "a world beyond physics." However, until the world beyond physics becomes part of the world within physics, science cannot rest. Only once the living world can be conceptually integrated within the physical sciences can the biological sciences finally sit self-assuredly at the Science High Table.[1]

The goal of the following chapters is to take steps towards that necessary integration by describing the recent discovery of a previously unrecognized kinetic dimension within chemical space. Material forms that occupy that kinetic dimension can be classified as *kinetic states of matter* in contradistinction to the established thermodynamic states of matter. Through the discovery of that kinetic dimension, an entirely new world of chemical possibility has opened up, one that nature has been occupying for almost four billion years, since the emergence of the first living forms. Significantly, the fact that all life forms exemplify kinetic states of matter allows living things to be placed within a coherent physical framework. As this text will describe in due course, that classification opens new doors to biological understanding. Finally, a more explicit physical/chemical answer to the "what is life" question may be emerging, thereby enabling life's remarkable nonphysical characteristics to be addressed in established physical/chemical terms. The physics-biology gap may be bridgeable after all.

2
Evolutionary Theory in Crisis

It hasn't gone to fisticuffs yet.

—Kevin Lala

The previous chapter described in some detail the difficulties faced by the physical sciences in seeking to place living things within a physicalist framework. Indeed, the inability of the physical sciences to come to terms with biological reality ended up having profound consequences on biology's conceptual development. Biology's theory of evolution—its centerpiece—is in essence a theory of change, but how can one formulate a theory of change when the physical basis of what is changing remains troublingly uncertain. That might help explain why evolutionary theory has been contentious for much of the past century. Let me then begin with a brief outline of evolutionary theory's path from its Darwinian beginnings and describe the conceptual crisis that developed over time.

Evolutionary Theory—A Brief History

Evolutionary theory has of course undergone substantial modifications since Charles Darwin first put forward his theory of natural selection. Significantly, his early evolutionary insights took place without any genetic knowledge. But with the discovery (or actually, rediscovery) of Mendelian genetics at the turn of the twentieth century, a new era for the subject opened up. By merging Darwin's theory of natural selection with Mendelian genetics, the neo-Darwinian synthesis, also termed the "modern synthesis," was formulated. Evolution was now understood to occur through random mutations in the genome's DNA, followed by natural selection acting on those DNA variants. Here was a theory of evolution that could be expressed in quantitative terms, enabling biology to proudly think of itself as a quantitative science, just like physics and chemistry, at least in central parts of it. And to cap it off, James Watson and Francis Crick's 1953 landmark discovery of the molecular structure of DNA cemented neo-Darwinian thinking as it provided unambiguous physical evidence for life's genetic basis.[1] The discovery of DNA transformed the concept of a gene from being merely conceptual to one that was physical—a segment of the extended DNA molecule with a clear functional role.

Biology's information era had begun. The genome was increasingly viewed as the set of instructions required to build a particular organism. In computer jargon, one might think of the genome as a computer program that, when run, carries out the program's directive. Each biological species is characterized by its particular program, as expressed by its particular DNA sequence. A mouse genome with its 2.8 billion nucleotide building blocks codes for making a mouse, while the *E. coli* bacterium, with just 4.6 million nucleotides, codes for the simplest existing life form, a bacterium. A human genome with its 3.2 billion nucleotides and some 20,000 protein coding genes would be considered a more complex life form than a relatively simple *E. coli* bacterium with just 4,300 genes.

All seemed to fit together perfectly. All that remained was to identify each of the genes and their specific function, then the entire recipe for life for that organism would be revealed. Evolutionary theory seemed to be on a roll and the powerful impact of Richard Dawkins's popularizing of the "selfish gene" concept further reinforced the view that evolutionary theory's golden age had arrived.[2] Biology's place at science's "high table," alongside physics and chemistry, seemed assured. The physical basis for life appeared to have been established.

Biology's Emerging Difficulties

But it was not long before developments in microbiology, the field of study dealing with bacteria, undermined established evolutionary thinking. It became increasingly clear that the simplicity of the Darwinian theme on which the modern synthesis was based—reproduction with random genetic variation—was incomplete. All biological systems are highly complex, and even bacteria, the simplest of life forms, manifest that complexity. The mechanisms of reproduction/replication turned out to be more elaborate than originally thought. Key biological processes that were discovered included horizontal gene transfer (the movement of genetic material *between* organisms, genetic drift (a process by which random events lead to genetic variation independently of natural selection), epigenetic effects (changes in gene expression and cellular function that do not involve alterations in the DNA sequence); and, finally, so-called Lamarckian inheritance (the heritable transmission of traits acquired by an organism during its lifetime). The idea that traits acquired during an organism's lifetime could be passed on to its offspring was first proposed by French biologist Jean-Baptiste Lamarck in 1801, but had been largely dismissed, even ridiculed, following the discovery of Mendelian genetics a century later.

The significance of these various mechanistic discoveries was troubling for evolutionary theory, as they seemed to undermine basic tenets of the modern synthesis. When University of St. Andrews evolutionary biologist Kevin Lala was asked at a 2016 Royal Society meeting, entitled "New Trends in Evolutionary Biology," how the conference was going, he responded: "I think it's going quite well. It hasn't gone to fisticuffs yet." The problematic issue that evolutionary biologists seemed unable to agree upon was whether the basic modern-synthesis framework formulated in mid-twentieth century was sufficiently flexible to encompass those new mechanistic discoveries. Some biologists claimed that the incorporation of these mechanistic details into the modern synthesis was possible without undermining the central evolutionary thesis of genetic variation and natural selection. Others considered that a more detailed theory, an *extended evolutionary synthesis* (EES), was necessary.[3] Significantly, however, life's essence as a biological phenomenon grounded in the genome's central governing role was not seriously questioned and continued to prevail.

Yet, even with these developments, theoretical difficulties at a deeper level continued to trouble. Life's most striking features—its purposeful, goal-directed, and cognitive nature; its highly complex structure—remained inexplicable in physical/chemical terms. Indeed, that disconnect between the physical and biological sciences only served to reinforce biology's "go it alone" path. Ernst Mayr, the evolutionary biologist who formalized the "autonomy of biology" approach, claimed that life's complexity and its evolutionary history meant biology could not be adequately understood through physical and chemical principles. He argued that biology is fundamentally a *historic* science, as it studies processes and patterns that unfold over time and are shaped by historical events. In his view, contingency plays a key role in governing biological outcomes necessitating biology's autonomous approach.[4]

Physical scientists approach their subject differently. They typically focus on the formulation of laws that are *time-independent,* such as Newton's laws of motion or the principles of thermodynamics. Such laws apply universally and are not tied to historical events. Physical/chemical processes can generally be reproduced under the same conditions, whereas biology is heavily impacted by its unique, unrepeatable history. As Stephen Jay Gould, the Harvard paleontologist, once remarked: if we could "rewind the tape of life" and were to play it again, evolution would take a very different course.

This fundamental difference between the physical and biological approaches led Mayr to argue that biological phenomena were inherently *irreducible* to physics and chemistry. It was the evolutionary process that created life's unique emergent properties—function, purpose, and goal-directedness—which, through natural selection, were optimized for survival and reproduction. So, though Mayr did not deny the appearance of purpose and goal-directedness in life activities, he argued that these characteristics were not *intrinsic* but rather were byproducts of natural selection. In his words: "The so-called purposiveness of living organisms is the result of a purely mechanical process, the result of natural selection."[5] Some years later, Richard Dawkins offered the same argument in *The Blind Watchmaker*: "Natural selection is the blind watchmaker, blind because it does not see ahead, does not plan consequences, has no purpose in view."[6] This view became the dominant theme in modern evolutionary biology.

Yet one could question that view. To claim that purposefulness (teleology) in life may seem real, but it isn't, that it is only "apparent," was a bold call. As will be discussed in chapter 9, the evidence for purpose within biology has become substantive. And if the basis for the claim is that the existence of purpose doesn't conform to established biological theory, that would not be sufficient. It could also mean that the theory itself requires modification, rather than the rejection of observations that question the theory. As Groucho Marx, the American comic, once put it: "Who ya gonna believe, me or your lyin' eyes?"

But before addressing further instances of emergent properties that remain unexplained in physical terms, let me briefly discuss the general question of emergent properties and the scientific methodology of reductionism by which one routinely attempts to explain such properties.

Reductionism in Science

Reductionism has been central to scientific methodology for centuries, with its roots in ancient Greek philosophy.

Its central idea is disarmingly simple: a complex entity can be understood through knowledge of its individual parts and how they interact. If you want to understand how a mechanical watch works, look at its component parts—cogs, springs, gears—and see how they fit together. Complex systems typically express "emergent properties," properties that are not found in any of the entity's component parts. Watches tell the time; their individual components are unable to do so.

In physics and chemistry, reductionist methodology has been exceptionally powerful. Much chemical understanding has come about through the ability to reduce chemical phenomena to underlying physical principles. The hardness of a diamond, the conductivity of metals, the softness of rubber—these are examples of emergent properties of matter, which are not present in the material's constituent parts. Happily, through reductionist reasoning, such properties are readily explained. A diamond's hardness derives from the strong carbon-carbon bonds connecting the diamond's carbon atoms. Metals conduct electricity because the outermost electrons of metal atoms are relatively free and can move throughout the metal's physical structure. The softness of rubber derives from the weak intermolecular forces that exist between the rubber molecules. In physics and chemistry, emergent properties are routinely explained in reductionist fashion.

In biology, reductionist thinking is no less extensive. Understanding the cell and its internal workings is fundamentally a reductionist enterprise. How do cells divide, how do enzymes catalyze chemical reactions, how do ribosomes synthesize proteins from amino acids, how do DNA molecules replicate and become two DNA molecules? How do neurons communicate? Our understanding of such issues is overwhelmingly reductionist. It was the dramatic advances in molecular biology over recent decades that provided the molecular detail enabling reductionist methodology to be applied in such cases. And, of course, understanding organismic function is also reductionist in its approach: you walk, talk and go about your daily affairs thanks to your kidneys, hearts, lungs, and so on, whose functions are based on a myriad of mechanistic processes, most of which are also understood through reductionist thinking.

But then a problem arises. There are basic global life characteristics, such as its purposeful nature and its mental capabilities, that emerged through the evolutionary process, and which are not explicable through reductionist methodology. Such properties are considered "irreducible." In those cases, *holism* is proposed as a counter-methodology to that of reductionism. Holism is the view that systems and their properties should be analyzed as *wholes*, not just as collections of parts. It emphasizes *emergence* and *interdependence*. The basic argument offered in such cases is that such properties arise from the complex interactions found in biological systems, which cannot be readily separated out.

But aspects of the holistic argument can be problematic. The difficulty is that in seeking explanations for life characteristics, holism offers no insight into the basis for those characteristics—a sleight-of-hand-type explanation. Even the workings of the brain, that extraordinarily complex organ, seemingly irreducible due to its complexity, have benefited from a degree of reductionist thinking. By studying the simpler nervous systems of the sea slug, Aplysia, the Nobel Prize-winning neuroscientist Eric Kandel uncovered the molecular basis for various neural mechanisms, such as short- and long-term memory, and helped bridge the gap between molecular biology and cognitive neuroscience.[7] Much of biological function is understood through reductionist methodology. Holism as an explanatory methodology can be unkindly thought of as an evasive way of saying "we still don't understand this." In his classic monograph "Chance and Necessity," Jacques Monod expressed that point of view rather bluntly: *"A most foolish and wrong-headed quarrel it is, merely testifying to the "holist" profound misappreciation of scientific method."*[8]

The bottom line is increasingly clear: until there is a coherent physical/chemical theory of life that can explain what life is, and how it was able to emerge from inanimate beginnings, it will not be possible to formulate a viable theory of evolution that is able to explain life's emergent properties. A central goal of this book is to outline such a theory, to hopefully enable many of life's striking characteristics, including purpose and mind, to be reduced to their physical/chemical beginnings.

Let me now describe several other life characteristics that traditional biological theory currently struggles to explain.

Life's Dynamic Nature

Structurally speaking, living things are strange, very strange. An animal today might look much the same as it will in some weeks or months, but it won't *be* the same. The organism's component bits and pieces have largely been replaced by other equivalent bits and pieces, very much like water drops turning over in a fountain. A fountain appears unchanged over time, even though the water that makes up the fountain is continually changing. What that means is that in a strict material sense, some months from now, that you, reading these pages, will effectively be an entirely different person! You may look the same, but you won't be. Let's take a closer look at life's constituent molecules to describe how that turnover process operates.

Consider cellular proteins, a key life component. Much of the functional machinery of an organism is composed of proteins, so their structures have to be meticulously maintained to fulfill those many functions. As a result, they can have surprisingly short lifetimes, sometimes measured in minutes. Those limited lifetimes are not due to their being intrinsically unstable, but rather because the cell continually degrades and re-synthesizes its proteins for optimal function. Damaged proteins need to be repaired, and the cell's requirements continually change at different stages of the cell cycle. Accordingly, cellular proteins are continually turning over through what is termed the "ubiquitin" mechanism. In that mechanism, once the cell "decides" that a protein is ready for degradation, it labels that protein with ubiquitin, a small protein, initiating the degradation process. Protein recycling is a crucial part of the cell's control and regulation mechanisms.[9]

Lipids, another vital molecule and major component of a cell's membrane, also undergo continual turnover, though at a slower rate. Such turnover is crucial for regulating material transport in and out of the cell, thereby maintaining the cell's homeostatic state. And as pointed out by John Mattick, the University of NSW biologist, similar considerations apply to the cell's many different RNA molecules.[10] Messenger RNAs typically have short half-lives ranging from minutes to several hours, allowing for rapid regulation of gene expression, while ribosomal RNAs and transfer RNAs also turn over, though with longer half-lives for functional reasons.

Interestingly, life's dynamic character even extends to the cell's DNA genome. The cell's genome does not undergo dynamic turnover like life's other building block molecules, but rather can undergo dynamic internal reorganization to satisfy the cell's needs at any moment. Barbara McClintock, the far-sighted American geneticist, in her study of genetic behavior in maize as part of work initiated in the 1940s, already recognized the genome's dynamic nature through her discovery of transposons, or "jumping genes."[11] The discovery was revolutionary at the time, as it questioned the established view that the genome was a static repository of information, rather than a flexible dynamic system capable of adapting to the cell's needs. Effectively cells were found able to *reprogram* themselves in response to changing conditions. Such unorthodox thinking was met with considerable skepticism at the time. Several decades later, researchers identified transposable elements in other organisms, confirming that transposons were not unique to maize but were widespread across different species. Belatedly, McClintock was awarded the 1983 Nobel Prize in Physiology or Medicine for her pioneering work.

More recent work by James Shapiro, an American biologist at the University of Chicago, has supported and extended McClintock's pioneering work. With his concept of natural genetic engineering, Shapiro claimed that the genome functions as a *read-write* system, reaffirming the genome's role as a dynamic, interactive system capable of modifying and reorganizing itself in response to internal and external signals.[12] That contrasts with the traditional *read-only* view, which holds that the genome acts as a fixed repository of information, operating as life's instruction manual.

The molecular picture that emerges is a striking one. Metaphorically, one might think of any living system as a *"chemical fountain"* with its component parts continually turning over, a *dynamic network* continually responding to environmental signals. It was that realization that led Barbara McClintock to claim that it is the cell that governs the genome, not the other way around. That dramatic statement changes our understanding of the genome's role in life processes and the nature of the evolutionary process.[13] The gene-centric view of life, and an evolutionary process based on that view, have become increasingly problematic, resulting in the dynamic view gaining increasing support in recent years. Of course, such dynamic behavior is in stark contrast to typical inanimate systems, say rocks or crystals, whose molecular and atomic constituents remain unchanged over time. The term "stable as a rock" means just what it says.

Interestingly, beyond the process of protein turnover discussed earlier, proteins have in some cases also been discovered to be structurally dynamic, very much like the cell's genome, and so able to take on different cellular tasks depending on the cell's requirements. For example, aconitase, an enzyme that operates in the citric acid cycle and is central to cellular respiration and energy production, can also act as an iron regulatory protein. When the cell finds itself under conditions of low iron availability, aconitase undergoes structural reorganization, enabling it to switch its function from catalysis to regulation. That plasticity in certain protein structures facilitates improved cell functionality. Yet here we are, half a century after McClintock's insights were recognized, and biology has still to accommodate this new dynamic way of thinking. It has become increasingly clear that the cell genome is not merely a passive instruction manual, but rather a dynamic and responsive participant in cellular processes. Much of the conceptual crisis biology has undergone during the past half century may be attributed to that central misunderstanding.

The Mind-Matter Divide

Finally, there is no clearer indication of evolutionary theory's conceptual difficulties than the unexplained existence of life's mental dimension. How did mind and consciousness emerge from "dead" matter? But they did. Let me briefly describe the establishment views on this ultimate life mystery.

In contrast to life's other emergent properties, the characteristic of mind is harder to marginalize by labelling it as "apparent." René Descartes, the seventeenth-century French philosopher, put his finger on it centuries ago—our mental existence is the only thing we can be quite sure of. But once the existence of both physical and mental worlds is acknowledged, we are led to ask how the two worlds are connected. Science readily explains the relationship among three physical states—solid, liquid, gas—but what is the relationship between physical and mental states? Science remains in the dark on this central question. Current evolutionary theory offers no physical insight into the nature of mind and how it emerged. So let me begin with some background.

In the latter part of the twentieth century, neuroscience studies invested considerable effort into understanding the mind phenomenon and its possible physical basis. Mind was considered an emergent phenomenon arising from the complex interactions of neurons and neural networks in the brain. An early assumption was that the brain could be thought of as a biological computer whose workings were analogous to that of an electronic computer, with the mind being the output of neural computation. But flaws in that computer analogy quickly emerged. It became increasingly apparent that the brain operates as a complex dynamic adaptive system, very much akin to the dynamic behavior found for cellular function discussed earlier, and so quite different from algorithmically based computer function. In any case, the analogy struggles to explain subjective experience (qualia)—the "hard problem" of how physical processes produce the feeling of being. Few would disagree with the statement that computers do not have subjective feelings.

But a dramatic development regarding the neural perspective on mind was yet to come. In the early part of this century, it became increasingly apparent that it does not take a brain or a neural system for mind-like characteristics to be possible. That understanding didn't just move the goal posts—it removed them altogether! Central to recent microbiological studies was the realization that nonneural life—simple bacteria, single-cell eukaryotes such as amoebas, even multi-cell life forms such as slime molds, creatures with no brain or neural system of any kind—can undertake complex decision-making and manifest adaptive, goal-directed behavior. That remarkable finding significantly transformed our understanding of cognitive function. It has shifted the concept of cognition from being a property exclusive to organisms with complex nervous systems to a fundamental property of life itself. Michael Levin's research has pointed out the key role of bioelectric signals in decision-making and behavior in nonneural systems, indicating that primitive "mental" processes predate nervous systems. As Levin puts it: nonneural intelligence exists.[14]

That discovery has profound implications. Earliest life did not just emerge from nowhere but presumably emerged in an earlier chemical phase along the evolutionary path. And that suggests that intelligence, cognition, and mind may well have their roots in chemistry. Mind in its most rudimentary expression may need to be thought of as *chemical*, not biological! I will explore where within chemistry the roots of cognition and mind might be found, but first we need to survey some more of the scientific terrain.

Evolution's Driving Force

Beyond those "big" life questions, more mundane, yet fundamental, life questions also await resolution. Does the evolutionary process have an identifiable driving force? And, if so, where does it direct? Can the force be expressed in physical terms? Evolutionary theory offers contradictory views on this question. Some biologists invoke natural selection as evolution's primary driving force. Others contend there is no driving force, that natural selection is the "blind watchmaker." Either way, the consensus view appears to be that evolution has no inherent direction.

But here also, uncertainties arise. The phenomenon of convergent evolution raises doubts regarding the suggested lack of direction. Convergent evolution occurs when unrelated species evolve similar traits independently, as observed in the evolution of wings in bats, birds, and insects or the streamlined shape of marine animals, as found in dolphins and sharks. Even the spiders' silken thread, spun to trap prey, protect their young, and more, is not unique to the *Arachnida* class (spiders, scorpions, and so on). Silk moths, fungus gnats and midges, and members of the *Insecta* class independently evolved to produce silk for similar functional reasons. Cambridge University paleontologist Simon Conway Morris has argued that convergent evolution implies a form of directionality, given that evolution repeatedly "discovers" in different species the same solution to environmental challenges.[15]

But there is more. The issue of increasing organismic complexity during evolution is also suggestive of directionality. Over time, life has evolved from simple single-celled organisms to highly complex multicellular organisms, and that complexification process has been interpreted as a kind of directionality or "progress" in evolution. Many biologists, however, argue that this increase in complexity is not driven by any inherent tendency toward complexity but, rather, is a consequence of selective pressures that simply favor complexity in some lineages (for example, animals and plants), while others prefer to remain simple (for example, bacteria). According to that way of thinking, the emergence of complexity could be viewed as an evolutionary consequence rather than as an evolutionary goal, with many organisms (such as bacteria) remaining highly successful in their simple unicellular mode. Stephen Jay Gould, who stressed the importance of contingency in evolution, supported that view.

And yet, the level of complexification that took place during the evolutionary process from inanimate to simplest life was extraordinary, staggering in fact, involving some nine or ten orders of magnitude (keep in mind that some ten billion molecules make up even the simplest life form). That's the difference between the size of a small coin and Planet Earth! Could that level of complexity have come about through Gould's random walk mechanism, namely by chance? It seems unlikely. And if natural selection is fingered as the likely suspect, the underlying question then becomes: *why* was increasing complexity selected?

Biology has been dealing with these polemical issues since the days of Darwin, with no clear resolution. Valiant attempts by traditional biologists to support the conventional wisdom continue, but troubling questions continue to arise. Evolutionary theory, biology's foundational theory, its crown theory, has become increasingly conflicted.

Evolution's Century of Confusion: A Brief Summary

Well over a century has passed since Darwin's theory of natural selection laid the groundwork for understanding the life phenomenon and the evolutionary process that underpins it. The modern synthesis, as well as its EES extension, revolutionized biology with the ability to explain the evolution process in genetic terms. And yet, that gene-centered focus has been unable to resolve central life questions: *why dynamic, why complexity, why purpose, why mind?*

With hindsight, the problem with the gene-centered approach has become clearer. The genome, that extended DNA-information-rich molecule, though crucially important for cellular function, is just a molecule, and no individual molecule, no matter how informative, can explain life's essence. Life's defining characteristic—its aliveness—does not lie within that DNA molecule. Life's cognitive character, its self-awareness, its ability to make decisions and act on its own behalf, even "think"—remain inexplicable in physical terms. In fact, how can *any* collection of molecules exhibit a mental dimension, a property so central to life? Such questions remain unanswered, and they point to a fundamental gap in our physical understanding of the life phenomenon. The standard "emergent property" response merely clarifies where the problem lies.

Science has certainly illuminated many of life's mechanisms—the function of the ribosome, the role of the genome as a repository of information, the process of cell division, and countless others. These discoveries resemble the discovery of individual cogs in a vast and intricate machine. Yet the secret of the machine itself, the overarching principles that bring these parts together into a coherent, coordinated, self-sustaining functional way, goes beyond discovering life's individual cogs. Life's secret remains hidden within the complex whole, and uncovering the physical basis for the existence of that whole remains the challenge. As noted earlier, evolutionary theory is a theory of change, but it is not possible to formulate a definitive theory of change, without a clearer understanding of what it is that is changing. That, in a nutshell, is evolutionary theory's main problem: we still do not have a definitive physical/chemical theory of life. Biology cannot adequately address the life phenomena when the physical sciences remain unable to incorporate living things within a physicalist framework.

A theory of evolution that rests almost entirely on genomic attributes has proved insufficient. Carl Woese, the iconic scientist most responsible for advancing the late-twentieth-century molecular approach to genomics and molecular phylogenetics, expressed the problem bluntly. In his 2004 article titled "A New Biology for a New Century," he claimed that biology had lost its way, that it had become mired in molecular biological detail without a strategy for addressing life's basic questions.[16]

In the next chapter, I will review a century of research into the unsolved origin-of-life problem. As will be discussed, the existence of such a process drops tantalizing hints regarding the physical basis of life. If nature was able to achieve that transformation, and both fossil and genomic evidence supports the view that it was, it means the transformation is feasible, discoverable, and, therefore, scientifically understandable.

3
The Origin of Life Enigma

> Any cell carries with it the experience of a billion years of experimentation.
>
> —Max Delbrück

Uncovering the origin of life on Earth would not only tell us how life emerged and its chemical mechanism, but it would also likely go a long way toward resolving the bigger underlying mystery—in physical/chemical terms, what life is. Did matter become "alive" gradually as it proceeded along the emergence process, or did a sharp transition from nonliving to living take place at some stage, after which biological evolution began to operate? In fact, that question isn't normally asked. So let me begin by outlining a brief history of the origin-of-life (OOL) problem to cover what is known and how we might begin to fill in at least some of the mechanistic gaps.

The overwhelming scientific consensus for many years has been that the formation of Earth and its accompanying solar system took place some 4.6 billion years ago, and that no life on Earth existed at the time of its formation.

Interestingly, just knowing there was an origin-of-life process is informative of itself, even without knowing its mechanistic details. It tells us that a process of emergence must have taken place, that living and nonliving material forms are connected in some way, and that uncovering the physical/chemical process that led to life's emergence would help identify the nature of the connection. So, just as knowing how a mechanical watch works helps to explain what a watch is, knowing how life emerged from nonlife would help in explaining what life is.

Another important consequence follows. Despite the prevalent view that many of life's emergent properties are not reducible to physics and chemistry, the fact that simplest life emerged from nonliving matter in a physical chemical process indicates that life's properties should be reducible, at least in principle. It's the mechanical watch argument again. You can only understand watch function in terms of its component parts once you know what the components parts are and how they interrelate. Similarly with life—knowledge of life's molecular components would be a necessary first step for achieving reductionist understanding, though the way in which those components interrelate would also be crucial. But more on that in due course.

A Brief History of the OOL Problem

The OOL question has been an active scientific problem for a little over a century, with both *ahistorical* and *historical* aspects. The ahistorical aspects derive from the universal physical and chemical principles that necessarily underpinned those early physical and chemical processes. The historical aspect derives from the fact that we are attempting to describe physical/chemical events that took place in the very distant past, several billion years ago. But how can such distant historical events be revealed? Historical events are normally uncovered by accessing the historical record and, for the biological segment of the evolutionary process, that record has indeed been bountiful. Paleontology, the branch of science dealing with fossil plants and animals, has provided rich historical information on the evolutionary process and was a major contributing source in the formulation of Darwin's theory of evolution. The sole diagram in Darwin's treatise, the drawing of a simple phylogenetic tree of life (reproduced in figure 3.1), was largely based on early paleontological studies involving the analysis of plant and animal fossils that had been preserved in rocks.

More recently, the reality of such a phylogenetic tree has been greatly strengthened and extended back in time through *molecular phylogenetics*. That approach enables comparison of different life forms through analysis of the sequence of certain conserved molecules, usually nucleic acids or proteins. Based on that approach, Carl Woese discovered that the living universe is more correctly divided into *three* domains—archaea, bacteria, and eukaryotes—rather than just prokaryotes (a simpler life form) and eukaryotes (a larger, more complex life form), as had been previously thought.[1]

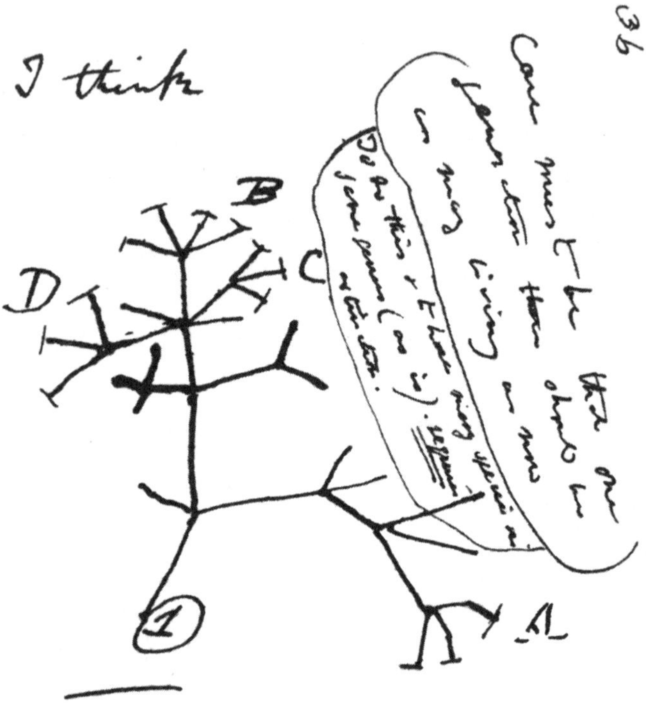

Figure 3.1
Tree of life diagram from Charles Darwin's *On the Origin of Species*, 1859.

But the molecular phylogenetic methodology, though undoubtedly powerful, also has its limitations. The molecular phylogenetic record reaches back roughly 3.8 billion years, farther than the physical paleontological record, but alas, it does not reach back to the actual origin of life phase. Thus, the very early record of life's emergence, much like the breadcrumbs in the story of Hansel and Gretel, has been lost in the mists of deep time. Phylogenetic tools are only useful for exploring the process of biological evolution, not the earlier phase of chemical evolution, the process by which simplest life emerged from inanimate matter.

The upshot is that if we want to reach back into deep time, closer to when life emerged, we need to employ alternative methods. Happily, such methods exist. As Max Delbrück, the Nobel-winning physicist-turned-biologist, put it: "any cell carries with it the experience of a billion years of experimentation." Each and every living cell on the planet today is descendant from earlier life, and that earlier life was itself descendant from the protocell entities from which life was able to emerge. Delbrück's message: hidden within every living cell are hints as to the process by which that cell came to be, its deep history. That is, if we know where to look.

But there is an *ahistorical* facet to the life question as well, and, as we will see, it is the more tractable one. After all, life is not considered to have come about entirely by chance. The probability of simplest life emerging by chance alone would have been infinitely small, so the process must have been governed by physical/chemical principles. What drove that physical-chemical process of emergence? It is the identification of those principles that continues to challenge the physical sciences.

The earliest scientific consideration of the OOL question was due to the independent contributions of Alexander Oparin, a Russian biochemist, and J.B.S. Haldane, a British evolutionary biologist, almost a century ago.

Their early pronouncements were groundbreaking for the period, first, because they argued for a material origin-of-life process, thereby rejecting vitalism, a belief still widespread at the turn of the twentieth century. Building on that material presumption, Oparin and Haldane postulated that a process of *chemical evolution* preceded a later process of biological evolution. That insight on its own was revolutionary, as it characterized the OOL problem as being one of *mechanism*. The assumption was that if the mechanism for the chemical evolutionary process could be uncovered, then that would constitute a solution to the OOL problem. Indeed, based on their knowledge of early colloid chemistry, they proposed that certain organic materials could undergo spontaneous self-assembly into what were termed *coacervates*—microscopic spherical aggregates of lipid molecules held together by various intermolecular forces. Such molecular aggregates are quite common in chemical thinking and are similar in size and appearance to the membranes that enclose living cells. It therefore seemed quite reasonable to propose such molecular aggregates as primordial cell-like structures. Molecular aggregates of that kind would be able to absorb and release components, merge, and even divide, thereby expressing certain life-like properties. Though this line of thought was not developed further at the time, the basis for a mechanistic approach to the OOL problem had become established.

Without doubt, however, the experiment that set the OOL field alight came some three decades later, when in 1952, Stanley Miller, a chemistry graduate student at the University of Chicago, carried out what was probably the most famous experiment in prebiotic chemistry.[2] The experiment was remarkably simple. Miller took a mixture of the gases thought at the time to have made up the atmosphere of prebiotic Earth—hydrogen, ammonia, methane, and water vapor—and subjected it to an electric discharge, thereby attempting to simulate the effect of primordial lightning. The result was dramatic and unexpected: the murky mixture that resulted was found to contain a range of amino acids, a key component of all living things. Proteins, which constitute life's structural building blocks, as well as the enzymes which allow life's chemical reactions to proceed, are made up of amino acids strung together in a long chain. The fact that amino acids, one of the key molecular building blocks of life, could be so readily formed from common materials under (presumed) prebiotic conditions seemed to signify in those heady days that resolution of the OOL mystery was imminent.

Following Miller's dramatic discovery, experiments to extend Miller's work were undertaken. The underlying logic was based on what could be called the *chemical/synthetic* approach to the problem. Given plausible prebiotic conditions, how could life's other key building blocks have formed? And once formed, how could their reaction and interaction lead to the emergence of a chemical system able to express simple proto-life character?

But a second approach, which could be termed the *physical/conceptual* one, was also taking shape. That approach involved identifying life's defining physical/chemical characteristics, primarily its heredity and metabolic capabilities, and attempting to identify the chemical means by which such characteristics could come about. Let us consider each of the two approaches in turn.

The Chemical/Synthetic Approach

Enthused by Miller's landmark experiment, chemists spent several decades exploring the prebiotic chemical landscape, seeking the chemical means by which other key biotic materials—lipids, sugars, and nucleotides (from which long chain nucleic acids are constructed)—could be formed. Though some of these reactions, as in the case of nucleotide synthesis, required quite creative chemical thinking, John Sutherland's group at Cambridge University was remarkably successful in pursuing that synthetic goal. Indeed, it was striking that within a few decades, feasible synthetic pathways for many of life's molecular building blocks, under presumed prebiotic conditions, were uncovered.[3] In retrospect, given the result of the Miller experiment, we should not have been too surprised by this synthetic success. The Miller experiment was extraordinary in the sense that amino acids were formed in a reaction that no chemist on Earth could have anticipated—the sparking of a mixture of very simple molecules, methane, water vapor, ammonia, and hydrogen, only one of which even contained a carbon atom, and a single one at that. Looking back, we might say that Miller's success at the time felt like someone buying their first ever lottery ticket and winning the million-dollar jackpot. No surprise then for the tremendous enthusiasm and optimism that ensued.

One of the less mentioned insights from the Miller experiment was what it revealed about the extent of our chemical knowledge. If such a spectacular and unexpected chemical result could be achieved through what we might consider a fortuitous chemical "shot in the dark," then how much more relevant chemistry is out there of which we are blissfully unaware? It seems safe to say that chemical potentiality greatly exceeds current chemical knowledge. And by a long shot! Nature very likely "discovered" many more feasible synthetic pathways toward life's building blocks than the limited ones that gifted chemical researchers happened to have uncovered over a limited period of several decades. To quote Arnold Schoenberg, the Austrian-American composer, in a quite different context: *"There is lots of music still to be written in C major."* While ignorance cannot be claimed as a scientific asset, awareness of that ignorance might in this case be informative. We should not underestimate nature's ability to surprise us—to carry out reactions whose existence we chemists are blissfully unaware of and could not have predicted. But does that mean that the origin-of-life question is primarily about scouring chemical space for undiscovered chemical reactions? My answer is a firm *no*. The solution, I believe, lies elsewhere. Let's see why.

Say you gather a group of the world's top synthetic chemists and biochemists and ask them to synthesize a simple living thing. For this synthetic exercise, all required chemical materials, reagents, and catalysts—whether simple or exotic, easy or hard to synthesize—would be supplied off the shelf, so knowledge of how those building blocks happened to appear some 3.5–4 billion years earlier would be less relevant. And creating the required reaction conditions for achieving that biotic synthesis? In principle, no problem there either. Any equipment able to create any desired reaction conditions would be supplied.

So how to begin? The materials chosen for the synthetic attempt would presumably be the biotic materials found within all living things—amino acids, nucleotides, sugars, lipids, and so on. But what reaction conditions should be applied? A reasonable choice would be to choose the reaction conditions that led to life's emergence close to four billion years ago. But what were they? Over the years, different locations and conditions have been proposed for life's emergence, and those proposals could not be more dissimilar. Darwin initiated the debate by suggesting that life began in some "warm little pond," but other locations have been suggested over the years. One proposal by Graham Cairns-Smith, a British chemist, which received considerable attention some forty years ago, hypothesized that life was initiated on a clay surface. The central claim was that primal replication was not based on organic entities but initially took place on a mineral surface. It was then suggested that more efficient organic-based replication emerged from that earlier mineral-based replication as an evolutionary adaptation. Creative, yes, but highly speculative, and I mention it here more for its historical interest.

A more recent proposal for life's primal location, which currently enjoys widespread support, is that of hydrothermal sea vents. These are effectively undersea volcanoes that exist in regions where volcanic activity is prevalent. Apart from these locations exhibiting wondrously diverse life forms, the vents are a rich source of reactive materials, primarily gaseous hydrogen and metal sulfide catalysts, which are postulated to have been able to reduce carbon dioxide to a variety of organic materials. Indeed, researchers have found that amino acids (the building blocks of proteins) can form under hydrothermal vent conditions, particularly in the presence of iron-sulfur minerals. But without going into mechanistic details, even here severe doubts have been voiced.[4] One researcher questioned whether the sodium-potassium ion concentration ratio in living systems is consistent with a sea water origin. Cell protoplasm is potassium-rich but sodium-poor, whereas in sea water the relative concentrations are reversed. Accordingly, it has been proposed more recently that the location for life's origins was more likely to be found in natural hot springs, rather than within sodium-rich sea water.[5]

With no more than that brief survey, the point should be clear. The picture is troublingly obscure—warm ponds, clay surfaces, hydrothermal vents, hot mineral springs. Not too much in common there. In fact, somewhat awkwardly, we seem to have come full circle. The location debate was initiated with Darwin's "warm little pond" (as proposed in his 1871 letter to a friend, Joseph Hooker) and 150 years later has ended up in a "hot spring"! The irony is hard to overlook. All proposals make interesting points, but none are falsifiable, a necessary condition for a scientific hypothesis to be constructive. What's more, no proposal offers insight into how life's building blocks, once formed prebiotically, might aggregate into the energized replicating chemical systems on the way to becoming "alive." Let's explore this point further.

Living things are chemically highly unusual—nonequilibrium chemical systems that maintain their nonequilibrium chemical state through continual supply of energy from an external source. But how could such a system have been able to emerge in the first place? Carrying out traditional chemical reactions in the hope that they will fortuitously lead us into this new dimension of chemical behavior by chance seems optimistic. Chemists have been carrying out a wide range of chemical reactions for several hundred years now, but the kind of chemistry that leads to life-like entities was not observed. A travel analogy may help explain why. Just as roaming endlessly on Earth's surface won't get you airborne and into that third spatial dimension, available evidence suggests that undertaking traditional chemical processes won't fortuitously get you into an energy-fueled, homeostatic nonequilibrium state. So, the problem at its core is more *conceptual* than *mechanistic*. Some out-of-the-box chemical thinking will be required to explain how access to that very special physical/chemical state, which seemingly characterizes all living things, was possible. Let us now consider the physical/conceptual approach in our search for a possible answer to the "why life" question.

The Physical/Conceptual Approach

In attempting to answer the "why life" question, we should first remind ourselves why *any* physical or chemical process occurs. As noted earlier, the answer to that question is unambiguous—it's that Second Law of Thermodynamics that we keep coming across, the law that explains why all physical and chemical transformations take place. But what law or laws could explain the emergence of a *nonequilibrium* chemical system able to maintain itself away from equilibrium over cosmic time scales? That is far from clear.

A second significant issue requiring explanation is the inordinate complexity of even the simplest living systems. Though mathematicians and physicists struggle to define complexity, as the famous quote goes, we know it when we see it. Life is incomprehensively complex. But the reason for that complexity remains obscure. Why did simple become complex? And to muddy the waters further, not all biologists agree with the general statement that complexity has increased over evolutionary time. After all, our planet continues to abound with bacteria, the simplest life form. No matter. Maybe the answer to the "why life" question can be found by answering the "why complexity" question. Can we somehow conceptually strip away the levels of complexity to reach some inner core, to uncover life's essence? Though life itself is complex, could it be that the life principle is relatively simple?

Gánti's Chemoton Model

Thinking along those lines, Tibor Gánti, a Hungarian theoretical biologist, composed the Chemoton model in the 1970s, an abstract model that sought to characterize minimal life.[6] In the model, he identified living things as being composed of three sub-systems integrated into a functional whole. His model did not mention specific molecular forms, but that was part of its appeal. It attempted to describe how any living thing would be constructed, just like an architect's sketches of a building would not need to specify the precise materials from which the building would be built. The three components that Gánti identified, though slightly modified over the years were: (a) a self-replicating informational system, (b) a metabolic system able to supply energy and chemical building blocks, and (c) a physical compartment to enclose the cell's components. That conceptual division opened new mechanistic horizons for experimental chemists. Reaching out for simplicity, chemists could now seek out chemical systems able to generate these capabilities separately. Indeed, through the discovery in the 1960s that certain molecules were able to self-replicate, a new direction in the study of the origin of life was initiated, one whose focus was on replication, in particular *molecular replication*.

Replication is central to the life process, and DNA replication, as part of that general process, is key, as it identifies the mechanism by which genetic information could be transmitted from one generation to the next. However, the process of DNA replication is a highly complex one involving many discrete steps, and requires a cellular environment for it to take place. But in a 1967 landmark experiment, Sol Spiegelman, a creative molecular biologist at the University of Illinois, carried out a molecular replication reaction on an RNA molecule, a nucleic acid cousin to DNA, and in a simple test-tube experiment, no less.

Spiegelman mixed an RNA strand (taken from the Qβ virus) with free-floating nucleotide building blocks and a catalyst enzyme.[7] The result was that the RNA molecule began making copies of itself. As illustrated schematically in figure 3.2a, RNA is a chain-like molecule composed of a series of nucleotide building blocks of four possible kinds, which can be labelled A, U, G, and C, and which appear in a specific sequence (for example: UCUUGAGCC). So how does the replicating RNA molecule manage to make an exact copy of itself from a mix of the four nucleotide building blocks, and in just the right sequence?

Replication takes place through the RNA molecule acting as a template. Free-floating nucleotides bind to complementary bases on the original RNA strand (figure 3.2b) in a lock-and-key manner, thereby ensuring an accurate sequence in the new strand being formed. The newly bound nucleotides then link together to form a complementary strand, creating a double-stranded molecule. Since the two strands are held together by weak bonds, they separate, resulting in the formation of *two* RNA molecules—the original and its complement. Of course, the second strand is not identical to the original—it can be thought of as a *negative* of the original. But in a second replication cycle, it is the complementary strand that serves as a template, producing a new strand (now a positive) that is identical to the original

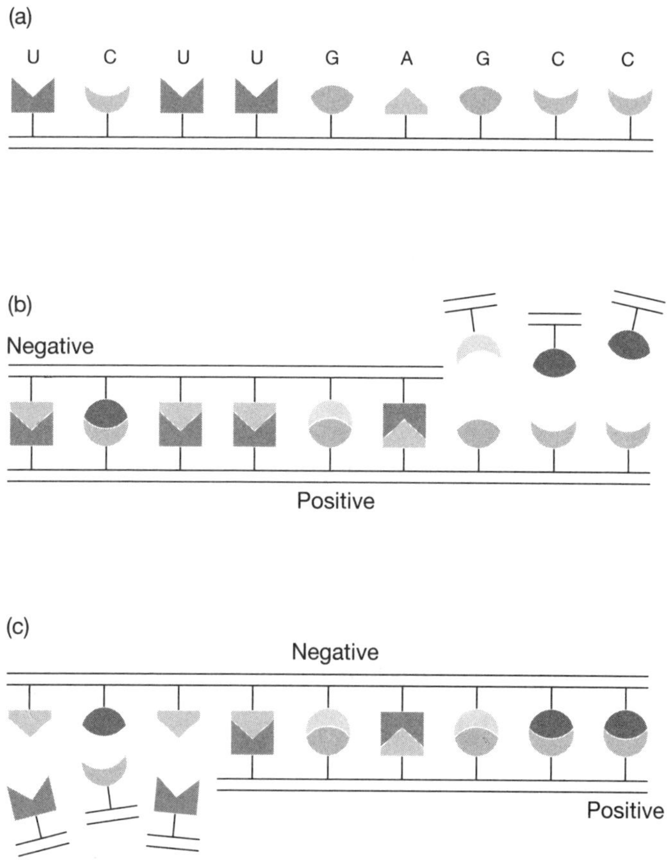

Figure 3.2
(a) Schematic representation of an RNA molecule made up from a sequence of nucleotide building blocks, A, U, G, C. (b) Representation of the process by which an RNA chain induces a complimentary copy of itself to be formed (positive to negative). (c) Representation of the process by which the complimentary RNA copy induces a copy of the original RNA to be formed (negative to positive). Source: Author created.

strand (figure 3.2c). Thus, after two template replication cycles, full self-replication is achieved.

But an even more dramatic result was yet to come. Spiegelman discovered that occasionally exact RNA replication did not take place. Sometimes copying errors occurred. Instead of a particular nucleotide, say G, fitting in at a certain position on the RNA template chain, another nucleotide, say A, would latch into place instead. As a result, the RNA copy ended up being slightly different from the original. In biological parlance, a "mutation" had occurred. And indeed, when that replication process was carried out multiple times, by continually transferring the product of replication from one test-tube to another, further mutations took place resulting in an evolutionary process. Indeed, after some seventy serial transfers, the original RNA molecule, some 4,500 nucleotides long, shortened to about 220 nucleotides.

The conclusion was striking: polymeric molecules with a sequenced structure, such as RNA, under the right conditions, were able to both replicate *and* evolve. Evolution was not just a biological phenomenon, but also a chemical one! Indeed, building on Spiegelman's landmark work, the three grand figures of OOL research at the time, Leslie Orgel, Francis Crick and Carl Woese, proposed the RNA-world theory for the origin of life.[8]

The essence of the RNA-world theory was that life began with RNA molecules, which were able to both store genetic information (like DNA) and catalyze chemical reactions (like proteins). This dual functionality suggested that at an early evolutionary stage, RNA could have supported simple self-replicating systems without the need for complex enzymes. One immediate benefit of the theory was that it appeared to resolve a long-standing OOL dilemma. Chemically speaking, all living things are based on a dual nucleic acid-protein system. Both structurally and functionally, life's exceptional operational capabilities derive from that integrated capability. But that dual nucleic acid–protein system creates a classic "chicken and egg" dilemma: for a cell to replicate and for the DNA genome within the cell to replicate as part of that general process, both nucleic acids *and* proteins are required. Without proteins, the cell's DNA cannot be replicated, and without the coded information written into the nucleic acid sequence, the required proteins cannot be formed. Neither can come into being without the other. So how could this dual system have come into existence?

The RNA-world theory offered a solution to the dilemma. It postulated that the more complex dual-DNA-protein world was preceded by a simpler world, in which RNA alone formed the basis of life. RNA, being more reactive than its DNA cousin, and being present in nature as a single strand polymer, might possess some catalytic activity. Thus, in that earlier world, RNA would be able to carry out both replicative *and* catalytic functions, though less effectively than the dual DNA-protein system. Indeed, in the early 1980s, Sidney Altman, from

Yale University, and Thomas Cech, from the University of Colorado, confirmed experimentally that nucleic acids were able to exhibit catalytic function, though, as anticipated, less effectively than proteins. Such RNA molecules were termed *ribozymes* to express both their RNA basis and their enzymatic capability.[9]

Reassuringly, following Max Delbrück's advice of looking carefully into cell structure to reveal the cell's history, support for the RNA-world theory was found hidden away inside. One key organelle, the ribosome, the exquisitely elaborate molecular machine which makes protein, is itself composed of multiple nucleic acid and protein molecules. But more careful inspection revealed an interesting feature of ribosome structure. The active core of the ribosome, where the function of generating proteins is carried out, is composed solely of RNA molecules. The protein molecules that are present in the ribosome are only found in the ribosome periphery, not in its active core. That observation led to the conclusion that early in evolution, the primal ribosome was an RNA construction, and that the incorporation of protein molecules into the structure was due to evolutionary "fine-tuning," which came about later. The existence of RNA's catalytic capabilities together with such historical information, as manifest in the ribosome structure, lent further support for the RNA-world theory.[10]

But that's where the good news ends. The RNA-world proposal runs into a deep problem. If the RNA-world thesis proposes that early life was based on RNA chemistry and on RNA's special chemical ability to undergo self-replication, then the study of the RNA replication reaction should shed light on the emergence-of-life process. The reality is that after more than half a century of study of RNA chemistry, following up on those early pioneering RNA works, no indication of an evolutionary process toward more life-like systems was observed. In fact, it was worse than that. As mentioned earlier, Spiegelman's RNA molecules, when subjected to continual rounds of replication, did show an evolutionary process, but toward faster-replicating, *shorter* RNAs, not toward more *complex* replicators—in other words, *away* from life, not toward it. Molecular evolution *was* directed, but the direction was toward the lifeless equilibrium state in accord with that ubiquitous Second Law. Clearly, some crucial element was missing. In the OOL context, yet another dead end.

Which leads us to the issue of metabolism, the second key component in Gánti's Chemoton model. Metabolic cycles are central to the functioning of all living things. Two well-known examples are the citric acid cycle, which leads to the oxidation of acetate to carbon dioxide, thereby providing the living entity with the energy needed to support life's systems, and the urea cycle, which converts toxic ammonia to less harmful urea. Such cycles are not chemical cycles in the usual sense, in that they are highly complex and are enzymatically controlled.

The question then arises: could *simple* nonenzymatic cycles have appeared spontaneously on prebiotic Earth as precursors to the complex cycles of extant life? But, in contrast to our discussion on replication, in which simple replication reactions are possible and have been observed experimentally, the emergence of cyclic chemical organization is more problematic.

For life to have emerged from such a cycle, it would have needed to be *autocatalytic,* meaning the cycle would have been able to make more of the components that make up the cycle. To illustrate, if there exists a set of molecules or molecular aggregates, A, B, C, D—such that A is able to catalyze the formation of B from available resources, and B is able to catalyze the formation of C from resources, which then catalyzes the formation of D, which catalyzes the formation of A, thereby closing the cycle—the cycle becomes autocatalytic, as illustrated schematically in figure 3.3. Catalytic closure means that the cycle catalyzes the formation of *all* its component parts. Thus, through autocatalysis the original set of A, B, C, D molecules becomes two sets of those molecules, which can react further to make additional sets.

In the origin of life context, the idea that life might have emerged from an autocatalytic cycle was first raised in the 1980s by Stuart Kauffman, the iconic theoretical biologist.[11] Such a chemical cycle at first might appear chemically implausible, but a moment's reflection indicates that the biological cell itself is effectively a holistic autocatalytic cycle. Cell replication does not

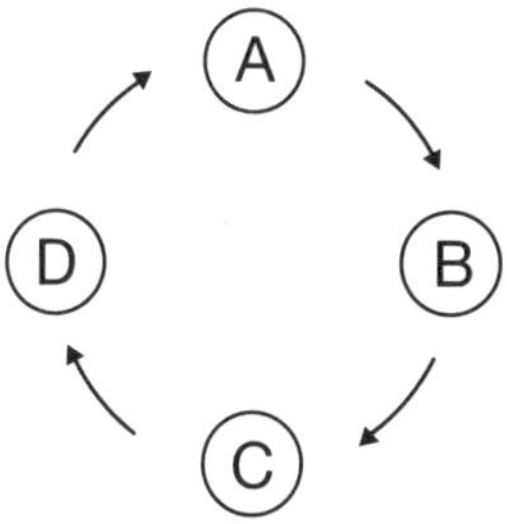

Figure 3.3
Schematic diagram showing the formation of an autocatalytic cycle composed of four chemical entities, A, B, C, and D, all expressing catalytic activity. The cycle is established through A catalyzing the formation of B, which catalyzes the formation of C, which in turn catalyzes the formation of D, which then catalyzes the formation of A, resulting in autocatalytic closure. Source: Author created.

proceed through each cell component copying itself, but rather through all the cell's components generating an autocatalytic set, resulting in a *single* cell becoming *two* cells—just as one autocatalytic set of A, B, C, D, molecules shown in figure 3.3, was able to become *two* sets of those molecules.

An appealing feature of the autocatalytic cycle proposal in the OOL context is that it eliminates the need to postulate the emergence of a relatively complex nucleic acid molecule with template-replicating potential to get the replication process up and running. The replicative capability of the cell does not lie within any *individual* member of the cycle, but within the cycle *as a whole*. Indeed, viewing the cell as a complex autocatalytic cycle offers possible insights into how life emerged through the fortuitous establishment of *simpler* autocatalytic cycles, which might then have complexified over time. In support of that way of thinking, certain metabolic processes, such as the reverse citric acid cycle, confirm the existence of such simpler autocatalytic cycles, as they can be found within existing life forms.

And yet, the autocatalytic cycle proposal for the origin of life leads to its own set of problems. First, outside of biology there is limited experimental support for the existence of autocatalytic chemical cycles, even simple ones. The best-known example of a chemical autocatalytic cycle is the so-called formose reaction, which involves the polymerization of a small organic molecule, formaldehyde. But the reaction is quite complex, leading to an inordinate mix of products and certainly shows no tendency to evolve toward a more coherent mixture. Other autocatalytic reactions have been uncovered over the years, but here also none appears able to evolve in a direction leading to the emergence of more complex assemblies with rudimentary life-like properties.

In summary, various approaches toward uncovering how nature was able to carry out its "magic trick" of transforming inanimate matter into living forms have been unable to resolve the OOL riddle. Two main theoretical contenders for explaining how complex life came to be, both the RNA-world theory and the autocatalytic cycle theory, run into the reality of Second Law constraints. The answer to the basic question as to how or why inanimate matter would be driven to complexify in a direction toward life has remained uncertain. Whether we start with a replicating molecule or an autocatalytic cycle, we run into the same Second Law dilemma. All chemical processes are broadly directed toward *equilibrium* and *disorganization*, not *nonequilibrium* and *complex organization*. Some central element remains missing.

The flying analogy mentioned earlier may be helpful in our attempts to characterize the dilemma. How does one get into the air and stay there? The Wright Brothers figured that out in 1903 by understanding the principles that allow heavier-than-air machines to become airborne. In seeking to solve the OOL riddle, we have effectively been scouring a two-dimensional chemical reaction surface, investigating every nook and cranny on that surface, seeking to find chemical systems expressing latent biological-like character, when the answer to the dilemma does not lie on that surface, but in a separate and unexplored chemical dimension. That other dimension involves *dynamic energized* chemical systems, where the rules of change are different and rudimentary biological properties can begin to be discerned. The existence of such a chemical dimension—a kinetic dimension—has only recently come to light, and it opens the door to a different *kind* of chemistry. The chemical nature of matter in that kinetic dimension, and its implications for both chemistry and biology, is the subject of Part Two.

II
Toward Resolution

4
Nature, the Ultimate Technologist

> If we examine the accomplishments of man in his most advanced endeavors, in theory and in practice, we find that the cell has done all this long before him, with greater resourcefulness and much greater efficiency.
>
> —Albert Claude

We humans pride ourselves on our technological capabilities. Starting off with simple applications, such as the use of stone tools by our early ancestors, humankind has made extraordinary technological advances. We now talk on phones, drive cars, and power our houses and factories with electricity. James Clerk Maxwell and Michael Faraday could not have imagined the impact their landmark ideas on electromagnetism would have on our everyday lives.

What is quite extraordinary, however, is that through the evolutionary process by which life emerged, nature also "discovered" and utilized technological capabilities.

But in the case of nature, it did so at a level of sophistication and efficiency that far exceeds those achieved by us humans, and did so millions, if not billions of years, before us. With electricity, for example, nature discovered its technological potential millions of years before Maxwell and Faraday put forward their theories. Electric eels, for example, can generate strong electric shocks (up to six hundred volts) to stun prey or defend themselves using specialized electric organs. Sharks, rays, and other aquatic animals can detect electric fields through specialized sensory organs that help them locate prey or navigate in murky environments. Plants are also into electricity utilization—for example, in their use of electrical signals to respond to stimuli, such as the Venus flytrap, which snaps shut when triggered. Electrical waves in plants can alert other parts of the plant to threats like herbivory, triggering chemical defenses. Muscle function in animals is caused by electrical signals that trigger the release of calcium ions, enabling muscle fibers to contract. In both plants (photosynthesis) and animals (cellular respiration), electrons are shuttled through proteins embedded in membranes, creating the energy gradients used to produce adenosine triphosphate (ATP), life's energy currency. Neurons communicate via electrical impulses, called action potentials, in which sodium and potassium ions move across membranes to generate voltage changes. Bioluminescence, the phenomenon whereby living organisms produce and emit light, is thought to have emerged some 400–500 million years ago.

But there's more, lots more. If you thought the information age began in the mid-twentieth century following the contributions of Alan Turing, John von Neumann, and Claude Shannon, it is enlightening to learn that nature discovered and exploited informational principles way before them.[1] The role of nucleic acid sequences in governing life processes reveals that nature was into digital information storage and processing principles billions of years ago, from the earliest days of life's emergence. Also, you might have been impressed with the recent discovery by Jennifer Doudna and Emmanuelle Charpentier of the revolutionary CRISPR gene-editing technology.[2] It allows scientists to edit DNA with unprecedented precision and efficiency with profound implications across different disciplines, especially for gene therapy applications. What is astonishing, however, is that nature discovered the CRISPR system billions of years earlier. Bacteria have been editing their DNA routinely for millennia, as part of their immune system. In fact, figuring out how bacteria do it was the basis for Doudna and Charpentier's Nobel Prize-winning discovery!

There are countless other examples of life's technological prowess. Take flight, for example. Nature exploited the technology of flight well before Swiss mathematician Daniel Bernoulli uncovered the principles governing flight in 1738. Insects took to the air millions of years earlier, having successfully applied the specific structural and dynamic requirements that stem from those aeronautical principles. In bees, the dynamics enabling flight turn out to be quite complicated. Additional forces, through wing rotation and vortex formation are required, as the short and stubby bee body cannot support flight through just wing beating. Even the common house fly exhibits remarkable technological prowess. They routinely land on a ceiling upside down, an exceptional aerobatic feat. In fact, human engineers have utilized high-speed videography to learn how that sophisticated maneuver is performed. Human engineers are more than happy to consult with house flies to advance their technological capabilities.

The list goes on. Nature has displayed staggering synthetic capabilities, having synthesized natural polymers—whether silk, wool, or cotton—eons before the chemical giant Dupont discovered nylon, the first synthetic fiber. And then there is the startling technological capability of the ribosome, that microscopic entity found in the thousands in every cell. That miniscule molecular machine is able to synthesize and churn out proteins in an assembly-line type process, utilizing the information coded into the cell's DNA in the span of a few seconds. The human synthetic chemist can only gaze in wonder at the extraordinary efficiency of that ribosomal system.

Without question, nature is the ultimate technologist—the supreme synthetic chemist, polymer scientist, electrical engineer, systems chemist, to name just a few of its wide-ranging capabilities. It is striking to realize that every living thing is continually exploiting nature's laws, and in a far more creative way than any human technologist, past or present. Paradoxically, in the case of simple life, it is doing so without those life forms even being aware that such laws exist.

But life's almost mystical capabilities do not stop there. There are examples of nature's technological capabilities that not only extend beyond human capability, but beyond current human comprehension. I am referring to consciousness, to mind. Nature discovered how to see—not just physically, as in sight, but conceptually, as in mind. Just as Einstein's theories of relativity added a new dimension to physical understanding, yet another physical dimension awaits discovery—the physical dimension of mind. In mind we have a natural capability that we have not yet begun to comprehend, let alone duplicate. In fact, it is a capability that seems awkwardly inconsistent with natural physical law, even though the existence of an evolutionary pathway by which it emerged indicates it must have been based on natural law, just like all the other technological innovations that nature "discovered."

So how did all this extraordinary ingenuity come about of its own accord? Charles Darwin provided us with the beginnings of an answer. Nature managed to access those technological capabilities, not by identifying the underlying physical laws, but by discovering their technological utility through an evolutionary process. But how did that process get started? How could natural selection have led to such unnatural consequences? The manner by which natural selection operates in purely chemical systems is well understood (in chemistry, natural selection is called "kinetic selection") but the outcome is very different. It results in chemical consequences—more of one product and less of another, a kind of zero-sum game. Natural selection operating on chemical systems has not yet been shown to induce the formation of biological systems, not even the hint of one. As described in the previous chapter, some seventy years of experimentation with molecular replicating chemical systems has shown no signs of an evolutionary process toward life-like systems.

Looking back, it is increasingly clear that nature's most extraordinary discovery was that of the biological cell—the simplest of life forms. Everything that nature discovered subsequently was little more than an afterthought. Alfred North Whitehead is often (mis-)quoted as saying that all philosophy is a series of footnotes to Plato. As we will discover, all life is a footnote to nature's ultimate discovery—the bacterial cell. That was nature's most ingenious discovery, one in which almost all of nature's creativity became expressed. In chapter 1, I discussed ontogeny, the process by which a single fertilized human egg, barely visible to the naked eye, divides continuously into something like 37 trillion cells of different types—heart, kidney, brain, skin, and so on—eventually being able to produce a Newton, an Einstein, able to invent mathematics, ponder nature's laws, and probe how the universe came to be. What law of nature could explain such an unfolding? Therein lies the deepest of scientific mysteries.

To understand life's unique features and how they emerged, we need to understand how nature's discovery of the cell came to be. But, as was discussed in chapter 3, the chemical evolutionary process that led to the emergence of that cell remains unknown. In the coming chapters we will consider recent developments in systems chemistry that led to the uncovering of basic laws of nature that, we believe, can help explain that extraordinary development.

To sum up, through an evolutionary process, nature "discovered" technology billions of years before humankind began to do so consciously. The implication appears unavoidable—the physical/chemical means by which chemical systems were able to become self-aware, to discover themselves, and the world outside, must lie within natural law. There lies the current challenge, to identify that law.

5
Simplifying Complexity

> We struggle with the complexities and avoid the simplicities.
>
> —Norman Vincent Peale

With all the uncertainty surrounding the nature of living things, there is one fact of which we are quite sure: even the simplest living entity, the bacterial cell, a speck of protoplasm one thousandth of a millimeter in width, is staggeringly complex. Despite its tiny dimensions, that cell consists of some ten billion molecules, many organized into complex functional structures that work together to maintain life. To put that number into perspective, the number of molecules in one single cell would be comparable to the number of components in over one thousand Boeing 747 airplanes. Unquestionably, life is complex. But how and why did that complexity come about?

A useful starting point might be to ask what complexity is and how it might be quantified. Physicists and mathematicians have grappled with that general question for some time.[1] Part of the challenge is that complexity is multi-faceted. Complexity in physics is not the same as complexity in biology, computational complexity is not the same as biological complexity, and information-theoretic complexity is not the same as organizational or functional complexity. In addressing the topic of complexity, context is key. The scientific dissonance was humorously captured by Conway Morris, who once remarked that the word "complexity" is akin to tasting a fine wine, with its nuanced descriptions: "forward oaty nose, thrilling undertones of blackberry, whiff of giraffe urine . . ."

Let me begin the discussion by briefly surveying the physical and biological perspectives, keeping in mind that our use of the term "complexity" will focus primarily on *organized functional complexity*—the integration of many interacting components into a coherent system capable of achieving specific outcomes.

The Physical View

Some seventy years ago, Erwin Schrödinger in his text *What Is Life?* puzzled over life's complex nature, though he phrased the question in an unusual way: "*Now, why are atoms so small? Clearly, the question is an evasion. For it is not really aimed at the size of the atoms. It is concerned with the size of organisms, more particularly with the size of our own corporeal selves.*"[2] In his thoughtful way, Schrödinger was directly addressing the complexity issue. He understood that life's complexity emerged from the enormous size difference between organisms compared to the molecular bits and pieces from which those organisms are constructed. So why are living things so complex? His tentative answer was to point to the randomness of atomic motion and the need to apply appropriate "statistical laws" to resolve that issue.

Actually, it was not by chance that Schrödinger sought a statistical explanation for resolving the life riddle. A statistical approach had been spectacularly successful in explaining the nature of change in the *nonliving* world, having become the basis for one of the most important laws in science, the Second Law of Thermodynamics, the subject of chapter 6. Although the Second Law does not forbid the local emergence of complexity in open systems maintained by energy flow, it offers no explanation for the sustained, functional complexity that characterizes living systems. The Second Law expectation would be that a physical system would be driven towards disorder and simplicity, not organization and complexity. In poker, you are less likely to be dealt a royal flush than a pair of sevens.

The Second Law, when applied simplistically, appears to argue against the emergence of complexity and life, and, if by some chance life did manage to fortuitously emerge, it would be expected to quickly decay back into nonliving irrelevance. Physical logic points us toward death and disorder, not to life and organization. Physics, by itself, has been unable to explain life's complexity.

The Biological View

In the biological context, as noted, complexity is of the functional organized kind, and manifests at every life level—molecular, cellular, organ, organismic, population. At the molecular level, key biological molecules—nucleic acids, proteins, lipids, carbohydrates—manifest complex structures that enable them to fulfill their prescribed functions. The ability of proteins, for example, to fold into precise three-dimensional forms is crucial for their enzymatic activity. Given that enzymes catalyze essentially all biological reactions, that structural complexity is integral to the functioning of every cell. RNA polymerase, for example, an enzyme responsible for transcribing DNA into RNA, a critical step in gene expression, is composed of twelve subunits with a hefty molecular weight of about 500kDa (kilodaltons). And then of course, there's the double helix structure of DNA, with its millions, and sometimes billions, of component nucleotide segments, which, by any standard, is complex. In humans, the DNA within a single cell is about two meters long when fully stretched out. No question, life's complexity is already evident at the molecular level.

At the cellular level, functional complexity continues to be the dominant theme. The cell contains within it a set of highly organized organelles, all with precise well-defined functions. The ribosome, for example, the machine-like structure present in all cells whose function is to synthesize the cell's proteins, is inordinately complex. With a molecular weight in the millions of Daltons, it "reads" the nucleotide sequence of the relevant RNA molecule and then proceeds to synthesize proteins whose amino building block sequences are based on the nucleotide sequences within those RNA molecules. Ribosomes are literally highly complex molecular machines in action. But other complex organelles are also integral to cellular structure—the nucleus, which acts as the cell's control center; the cell membrane, which serves as a protective barrier for the cell; the cell's mitochondria, which play a critical role in energy production, to mention key ones. Those various organelles are further expressions of the cell's structural complexity.

But life's complexity doesn't stop at the cell level. Approximately 1.2 billion years ago, multicellular organisms—you, me, and all other animals—emerged from single-cell creatures. Unicellular eukaryotes began to form multicellular aggregates, which then proceeded to undergo cell specialization and division of labor. Thus, each animal is comprised of organs—hearts, lungs, kidneys, and so on—each constituted from specialized cells, ultimately leading to the complex multicellular organisms we see today.

And finally, there are life's networks, yet another layer of functional complexity. Life forms do not function as individuals but connect to other life forms through an extensive array of networks. We can say life is fundamentally a network phenomenon. Bacteria, life's simplest individual form, though often perceived as solitary organisms, operate in extended networks. Individual bacteria communicate with one another through what is termed "quorum sensing"—a chemical communication mechanism utilized by bacteria to coordinate group behavior. That mechanism enables them to cooperate and adapt to their environments.[3] For example, bacteria, like many life forms, seek the protection of housing. The bacterial generation of biofilms, an extracellular matrix that effectively houses the bacteria, is just one example of such coordinated behavior.

But coordinated bacterial behavior goes further. Bacteria exchange genetic material through horizontal gene transfer and other processes to allow for the rapid spread of genes, thereby enabling the population as a whole to benefit from favorable genetic capabilities. Some bacteria exhibit collective swarming motility enabling them to explore new environments and access new sources of nutrients. The rich social networking of bacteria has been wonderfully detailed in recent years, revealing that such networking is not just restricted to intraspecies activity. Bacteria routinely engage in symbiotic relations with other organisms. Thus, nitrogen-fixing bacteria establish a mutualistic relationship with plant roots, which provide the plants with nitrogen, a crucial plant ingredient, and in return, receive nutrients that are essential for the bacteria's growth and energy needs. Closer to home, gut bacteria engage in a synergistic relationship with us, their human hosts. Gut bacteria help break down complex carbohydrates and other compounds that the human body cannot readily digest. They synthesize essential vitamins and even play a crucial role in the regulation of our immune system.

But in life there are no free lunches. In return for their beneficial services, we humans provide "our" bacteria with a nutrient-rich, stable, protective environment in which to thrive. Win-win! Life is anything but a zero-sum game. Thus, the biological view on complexity is uncontroversial: life's complexity arises from the interaction of variation, natural selection, and cooperation, operating over very long timescales.

But does the reality of life's complexity mean that the evolutionary process is directed toward increasing complexity? Here the biological view is less clear. The prevailing view is that evolution is not "about" increasing complexity, but, rather, is fundamentally about differential survival and reproduction in specific environments. Thus, complexity can increase as a by-product of that process, but it can also remain stable or even decrease. This is why many biologists resist the claim that evolution has an inherent trajectory toward complexification. There are clear cases of evolutionary simplification called "regressive evolution." Cave fish, for example, have lost their eyesight because vision confers no advantage in perpetual darkness and would be metabolically costly. Similar reductions in complexity are seen in many parasitic organisms that shed structures they no longer require. These examples demonstrate that complexity is not universally favored and that evolution can move in either direction depending on ecological context.

One influential attempt to explain the long-term pattern of increasing complexity without invoking a directional force was Stephen Jay Gould's "drunken walk" model.[4] He proposed that evolution resembles a random walk bounded by a wall of minimal complexity: once life begins at a basic level of organization, random variation alone will, over time, tend to generate lineages that drift toward higher complexity simply because they are bound by the wall's lower limit. On this view, complexification is not driven by a special biological tendency but emerges naturally from the structure of evolutionary possibility.

Yet the Gould explanation increasingly appears inadequate. First, it treats complexity primarily as an individual-organism property, whereas much of life's complexity is inherently relational and networked. As noted, "simple" bacteria are embedded in highly coordinated populations that communicate, exchange genes, and act collectively through mechanisms such as quorum sensing and biofilm formation. Once this social and ecological dimension is taken into account, the baseline from which evolution operates is already far more complex than Gould's model implies.

Second, the most significant historical increases in biological complexity—most notably the emergence of eukaryotic cells and multicellular life—were not the result of random drift away from a wall but resulted through what is termed symbiogenesis: a deep form of cooperation and integration between simpler prokaryotic organisms. Additionally, the coordination of specialized cells in multicellular organisms and the formation of interdependent ecological networks both suggest that complexification has been closely tied to the progressive layering of cooperative relationships rather than to mere statistical wandering.

When this is combined with the fact that the origin of life itself could not plausibly have involved the one-step random assembly of a fully functional cell, it becomes difficult to treat complexification as merely incidental. Even the simplest living systems exhibit an extraordinary level of organized complexity, one that far exceeds what would be expected from unguided molecular reorganization alone. Life's emergence therefore points not to sheer randomness, but to the operation of underlying physical processes capable of generating and sustaining organized structure.

Taken together, these considerations suggest that, over deep time, complexification has been a recurring and foundational feature of the living world, most significantly, through the establishment of increasingly extensive cooperative and symbiotic networks. The central question, then, is not whether complexity has increased over evolutionary time. It has become increasingly clear that it has—but *why*?

The Technological View

Insight into the basis for complexity in the biological world can be gleaned by observing how complexity manifests in the technological world. Take any technological device—a car, computer, or phone—and you'll see that new models become increasingly complex over time. The Wright Brothers' 1903 airplane evolved into the Boeing 747 with six million parts, and the early digital computer ENIAC, with its 18,000 vacuum tubes, was transformed into modern computers with more than 10 billion transistors. In the technological world, increased complexity takes place because increasing complexity invariably leads to improved *function*. Or put another way, it was the drive toward improved function that led to a complexification process. No surprise then that in the technological world we find ourselves surrounded by increasingly complex technological devices.

In the biological world, it seems that nature, the ultimate technologist, has, through an evolutionary process, exploited the same guiding principle. Through the action of natural selection, nature "discovered" the complexity-function relationship and proceeded to exploit it to the hilt, doing so at every operational level, whether molecular, organismic, or any level in between. Living things are complex for the same reason that phones, computers, and airplanes are complex—for improved functionality. Just as mobile phones became more complex with the release of each subsequent model, life's "newer models," obtained through natural selection, have also tended to be more complex. Functionality is central to all life's activities, and, accordingly, we can conclude that biological complexity evolved for the *functional* advantages it conferred.

Yet, revealing the function–complexity relationship does not entirely resolve the complexity dilemma. The difficulty is that the function–complexity relationship is problematic from a physicalist standpoint. Function, in its biological context, is inherently teleological (purpose-like) and does not arise naturally within the vocabulary of physical law. How did function arise? In the inanimate world, there is no function. The means by which the functionality-complexity relationship can be physicalized, to a degree at least, is central to illuminating life's greatest mystery. More on that question in chapter 8.

6
Stability: The Role of Energy

> We absolutely must leave room for doubt or there is no progress and there is no learning.
>
> —Richard Feynman

From the moment we become aware that there is a world out there, we discover the world is characterized by change. The Greek philosopher Heraclitus put it perceptively some 2,500 years ago: *there is nothing permanent except change.* Things happen, constantly. Change is the norm, change is everywhere. Before discussing the process of change in the life context, let me outline the central physical principle governing change throughout the universe—the Second Law of Thermodynamics.

Why Change? The Ubiquitous Second Law of Thermodynamics

The Second Law of Thermodynamics is one of nature's central laws. It governs all physical and chemical processes—why wood burns, why water freezes—but it also applies to all biological processes, including those metabolic reactions that maintain the living state. In fact, there is no physical or chemical process that can escape its rigid, ironclad demands. It is no less than extraordinary that one single law can be so absolute, so universal. But as was pointed out a little over a century ago by Ludwig Boltzmann, the great Austrian physicist, there is deep logic underpinning the Second Law—the logic of numbers, a logic that forms the basis for casino operation everywhere. (As an aside, his insight was so perceptive and far-reaching that his physicist colleagues at the time hesitated in accepting it. Boltzmann died (by committing suicide) without the brilliance of his insight having been fully accepted by the science of the day.) Let me then begin with a simplified theoretical description of the law and the logic that underpins it.

Due to its fundamental nature, the Second Law can be expressed in several seemingly unrelated ways. One common statement is that heat can only flow spontaneously from a hot body to a cold one, not in reverse. Hot coffee becomes cold, but cold coffee never becomes hot, even though in both cases the law of conservation of energy would be upheld. But that description hardly captures its deep significance, certainly not at first sight. A more intuitive description is that disorder is the norm—that order, when it exists, is transient and will ultimately revert to disorder. The physical concept of entropy, which can be thought of as a measure of the degree of disorder, will be discussed shortly.

The Second Law can also be expressed in energetic terms (actually, in terms of so-called "free energy," a particular kind of energy), namely, that matter in an *unstable* state, a so-called *nonequilibrium state*, will tend to change toward a more stable state, toward its *equilibrium state*. Add a drop of ink to a glass of water and you'll see the process directly. Initially an unstable state is created when the ink drop remains concentrated in one region of the glass. That's the more ordered state, the low entropy state. But before long, the ink spreads itself uniformly throughout the water and so generates a less ordered state. The initial unstable nonequilibrium state becomes transformed into a stable equilibrium state, one whose entropy is higher. Importantly, once that stable equilibrium state is reached, no matter how long one waits, the ink drop will never revert to its original state in which the ink drop was concentrated in one region of the glass. As there is no further change in the system's state, we can say that the final state *persists*.

The ink drop experiment is a simple one and may initially seem to be unrelated to ice melting, wood burning, or a sun dying. But that explicit direction of change, as reflected in the ink drop experiment, is true for every process in nature, physical or chemical, whether here on Earth or deep within the cosmos. In all cases, a more ordered state is transformed into a less ordered one. But what makes the Second Law so special is that though its initial formulation was *empirical*, meaning based on observation, Boltzmann's contribution showed its basis was more fundamental than that—that its basis was *logical, mathematical*.

Most physical laws, such as the law of gravity, for example, are empirical. The legendary Isaac Newton formulated that law in 1687 based on countless observations—falling apples, planetary motion, and the like. The Second Law of Thermodynamics, however, is valid for the same reason that one plus one equals two. We cannot imagine a world in which it would be inapplicable. That's why the Second Law is such a profoundly important law. Let me demonstrate that the Second Law is logical and grounded in mathematics.

Consider the following simple thought experiment. Six gas molecules are placed into two connected vessels, A and B, and the question arises as to how those six molecules will tend to distribute themselves between the two vessels? There are four possibilities: Will it be 3–3, 4–2, 5–1, or 6–0? Based on our experience with the tendency of gases to expand, we know that when a full gas cylinder is connected to an empty cylinder, the gas in the full cylinder will flow into the empty cylinder until the pressures in the two cylinders equalize. That's empirical. But why is that? Each molecule of gas moves about independently of all others (this is not entirely true, but its assumption does not affect the discussion), so why do the gas molecules partition themselves more or less equally between the two cylinders? Restricting our example to just six molecules, and assuming the molecules do move about independently, makes the explanation easy to understand. Logic alone shows that the likelihood of the six molecules dividing themselves 3–3 between the two vessels is *ten* times more likely than the possibility of all six molecules being found in just one of the two vessels. Let's see why.

The analysis begins by numbering each of the gas molecules—from 1 through to 6. We then ask how many ways there are of placing all six molecules in just one of the two vessels. The answer is clearly *two*; they may be all placed in Vessel A or all in Vessel B. No other possibilities exist. Now, we ask how many ways are there of dividing the six molecules *between* the two vessels, with three in Vessel A and three in Vessel B? Here, the possibilities are more numerous and it is easy to show that there are *twenty* distinct ways of dividing the six molecules *between* the two vessels. In other words, assuming that all the molecules are moving between the two vessels randomly and independently, the *chance* of finding the molecules divided 3–3 is *ten* times more likely than finding them all in one of the vessels (twenty ways vs. two ways). The laws of probability, entirely mathematical in nature, dictate that reality.

So much for the case of just six molecules. However, as the number of molecules increases, the probability of more equal distributions increases dramatically. If you start off with one hundred molecules, then the probability of finding the molecules divided between the two vessels, 50–50, as opposed to all hundred being in one vessel, jumps to 50,000,000,000,000,000,000,000,000,000 (that's 5.10^{28}) times more likely! Needless to say, if the number of molecules is more like the actual numbers that we would encounter in a real gas experiment, something like 10^{23} molecules, then the likelihood of all the molecules being in one vessel reaches vanishingly small numbers. That's why gases always expand throughout the space that is available; it does so for *probabilistic* reasons.

Now you may ask: what does this have to do with the direction of chemical reactions, say the burning of gasoline to produce carbon dioxide and water, or the many biochemical reactions that partake in a living organism's metabolic processes, or the expectation that our sun will become extinguished in about five billion years?

The answer is everything!

Consider a general chemical reaction in which two reactants, A and B, react to form two products, C and D, and which can also proceed in the reverse direction, from C and D back to A and B, as depicted below:

$$A + B \rightleftharpoons C + D$$

What will determine where the equilibrium will lie? Will it be toward reactants A and B, or toward products C and D, or somewhere in the middle? Every reaction has its preferred equilibrium state. For any chemical system, it turns out that the equilibrium concentrations, like with the gas molecules, also depends on probabilities—the number of molecular arrangements (called microstates) that are possible for A and B, versus the number that are possible for C and D. If the number of microstates describing A and B (meaning the number of ways A and B can be described) is greater than the number of microstates describing C and D, say by a factor of 10 to 1, then the equilibrium mixture will contain ten times as much of A and B than of C and D. Again, it's just that numbers game, with matter seeking to occupy its most probable state. That probability principle applies to *all* macroscopic change in the universe. The entropy concept is just the quantification of that basic idea. The more probable the particular state of a physical or chemical system, the greater its entropy. Boltzmann's landmark contribution was to propose a simple mathematical expression for entropy based on such probabilistic considerations:

$$S = k \ln W$$

where the entropy, S (simplistically, the degree of disorder of a system), is related to the number of microstates, W, that contribute to the system's description. The greater the number of possible microstates, the larger the entropy S (related via a constant, k, termed the Boltzmann constant). Thus, the degree of disorder and the number of possible microstates that describe a system and its entropy are mathematically connected. That is why the entropy of any system is driven toward its highest value. That's the state that is the most disordered, the one described by the largest number of microstates, and therefore the most probable.

The entropy concept is a powerful one, encompassing the nature of change all around us. It explains why shuffling an ordered deck of cards leads to a disordered deck, but shuffling a disordered deck will almost never get you back to an ordered deck. It explains why gases expand to fill their container. It explains why making hard-boiled eggs from raw eggs is easy, but turning hard-boiled eggs back into raw eggs is not possible. It explains why our sun will burn out in five billion years. It is fitting that the Boltzmann landmark entropy equation, one which encapsulates the logic of change throughout the universe, one his contemporaries struggled to accept, is engraved for posterity on his tomb in Vienna's Central Cemetery.

Returning to the life issue, you might at this point reasonably conclude that the origin-of-life question has therefore been solved. Why did life emerge? Because of the Second Law! If all transformations are governed by that Second Law, then the transformation from inanimate to animate must also be explained in that way. Indeed, there is a school of thought that sees life primarily as a thermodynamic phenomenon, proceeding in a "downhill" direction, like balls rolling down an incline. Nonetheless, despite the all-encompassing Second Law, the living state behaves in a way that is strange, seemingly counter to that logical directive. As I will now describe, not all physical and chemical phenomena can be adequately explained through a thermodynamic perspective.

At first glance, living things appear to disobey the Second Law. They maintain themselves in a highly organized, high-energy state even though the Second Law directive informs us that such a state should quickly degrade to a lower energy state. But they don't. They maintain a nonequilibrium state, characterized as *homeostatic*. As the nineteenth-century physiologist Claude Bernard, who discovered homeostasis, put it: "all the vital mechanisms, varied as they are, have only one object, that of preserving constant the conditions of life." The list of homeostatic physiological conditions that are maintained is impressively long. An obvious example is the way our body temperature is carefully maintained at close to 37 degrees Celsius regardless of the ambient temperature. But if you have ever taken a blood test to measure blood concentrations of key physiological parameters, you will know the list of monitored items is extensive. None of those measured concentrations are equilibrium concentrations in the chemical sense. The body maintains them all at nonequilibrium concentrations, at a level required for proper physiological function. That's homeostasis in action. If body chemistry were to seriously deviate from its homeostatically prescribed levels, it would mean death, something all living things try hard to avoid.

So how is a nonequilibrium state able to be maintained over time at all? Wouldn't that constitute a breakdown of the Second Law? The answer is *no*. An unstable, nonequilibrium state, able to persist over time, *is* possible provided a source of energy is continually supplied to the system. The way this can happen is readily illustrated with the physical example of a fountain (figure 6.1). The Jet d'Eau on Lake Geneva is an impressive fountain able to pump water at a rate of 500 liters per second and results in a water jet reaching a height of 140 meters. At first sight, that fountain seemingly defies the law of gravity, as what appears before you is a body of water—the water jet—suspended in space over extended time. That structure is unstable energetically, yet it is maintained. Of course, there is no mystery here. The water in the water jet is not the same water from moment to moment but is continually changing. The physical nonequilibrium steady state that constitutes a fountain can be maintained indefinitely provided the pump that operates the fountain is continually supplied with water and energy.

Well, believe it or not, in certain respects, the living state operates in much the same way as the fountain. Any living state is also a nonequilibrium dynamic steady-state system, which is maintained over time, provided a continual supply of energy and resources is available. Indeed, metaphorically speaking, every living thing could be thought of as a *chemical fountain*. The process that takes place within that dynamic system, both its formation and its decay, is chemical, not physical. In living things, the energy is obtained either

Figure 6.1
Photo of the Jet d'Eau in Geneva, Switzerland, depicting a body of water in a dynamic nonequilibrium state. Source: Photo taken by author's son.

from a chemical source, the food we consume, or, in the case of plants, algae, and certain bacteria, from a photochemical source, sunlight. So living things, though they are nonequilibrium systems, fully obey the Second Law of Thermodynamics, just like fountains. On this point at least, biology and physics are in harmony. The nonequilibrium dynamic steady-state that characterizes living things can be maintained over time through energy-producing metabolic processes. That insight was a central conclusion for Schrödinger back in 1944.

But that pattern of thermodynamic behavior in living things is nonetheless strange. The Jet d'Eau fountain in Geneva maintains its steady state through human design, not through natural causes. In the case of living things, the maintenance of that high-energy homeostatic state through deliberate actions came about of its own accord and is physically unexpected. How could such a dynamic state have emerged naturally? Despite considerable variability in the food we eat and the activities we undertake, our body and blood compositions undergo little change. Attempt to knock one of the system's key parameters out of its comfort zone and a set of feedback mechanisms immediately kick in and the system reverts to its preferred steady-state. Of course, if the perturbation on the system is too persistent or too large, then the entire holistic entity can collapse, causing the living thing to die. But what struck Bernard was the seemingly holistic and irreducible nature of the "life machine." The deliberate holistic nature of living things, as expressed through homeostasis, is a clear mechanistic expression of *agency*, the capacity of a system to act to maintain itself. Homeostasis, as a manifestation of agency, was sufficiently puzzling that it led during Bernard's time to widespread support for a vitalist view on life, one that though it is now scientifically no longer credited, continues to be promulgated by advocates of Intelligent Design.

While modern science emphatically denies that vitalist position, the question remains: what is the underlying physical/chemical logic for the establishment of that chemically unstable homeostatic state? How did the emergence of an intricate and highly organized energy-fueled homeostatic state, able to accommodate the Second Law's strict requirements, come about in the first place? It was those kinds of concerns that caused the leading twentieth-century physicists—Niels Bohr, Erwin Schrödinger, and Eugene Wigner, mentioned earlier—to express perplexity over the life issue. To put it bluntly, to a physicist, the very existence of a biological world, expressing purposeful, intentional, cognitive behavior, does not make physical sense. As described in chapter 1, physical and biological logic are seemingly incompatible. And biology's attempt to explain away this issue by simply labelling biology's extraordinary functionality as the expected outcome of natural selection operating over geological time is inadequate. Natural selection broadly explains how life, once it exists, was able to evolve. But it is unable to explain the origin of life—how "dead" stuff became alive, how a chemical system was transformed into a biological one. What is the chemical basis of life's "aliveness?" The homeostatic state does not violate the Second Law in letter, but it certainly violates it in spirit.

The Path from Instability to Stability

A further advance in our understanding of the life phenomenon took place in the 1970s through the work of Ilya Prigogine. His contribution was to a relatively new area, which goes under the fancy title of *nonequilibrium thermodynamics*, and won him the 1977 Nobel Prize in Chemistry.

As described earlier, a physical-chemical system that is unstable strives to become stable by reaching its equilibrium state, as directed by the Second Law. But in doing so, it often displays quite unusual characteristics. Let's illustrate this with a simple physical example—the familiar vortex that appears in a bath of water once the plug is pulled. The act of pulling the plug converts a *stable equilibrium system* into an *unstable nonequilibrium* one. Accordingly, the body of water seeks to lower its energy to restore it to an equilibrium state. The key point is that in seeking to reach that lower energy state, it creates the familiar vortex—a relatively ordered *unstable nonequilibrium* structure that is maintained as long as that nonequilibrium state is maintained. The reason for the emergence of this transient structure is to speed up water flow and restore the body of water to an equilibrium state. That ordered structure is termed a *"dissipative structure"* because a source of energy that maintains the structure has to be continually dissipated for it to be maintained. That dissipative structure phenomenon goes far beyond emptying baths. A satellite picture of a hurricane shows the same structure, but in that case, it may well extend over hundreds of kilometers. In the case of hurricanes, the nonequilibrium state arises from the difference in pressure between two atmospheric regions, which is dissipated by the same dissipative structure mechanism observed for an emptying bath.

So what is the significance of dissipative structures to an understanding of the life phenomenon? Prigogine proposed that living things could themselves be thought of as dissipative structures. In that view, life as a persistent nonequilibrium entity emerged as a means of dissipating energy, whether chemical (through the food we eat) or photochemical in the case of photosynthetic life (plants and certain bacteria). The argument proposed was that life is yet another way that nature's operative Second Law has come up with to dissipate energy, to increase entropy.

But attempting to relate physical dissipative structure with living things is now thought of as a bit of a stretch. There is more to life than just the response of a physical system to the effects of an energy gradient and the drive toward energy dissipation. If that were the case, one might expect living forms to spontaneously emerge in the vicinity of energy gradients, just like vortices, but they don't. In any case, it would be difficult to consider Pavarotti singing *O Sole Mio* to a packed concert hall as merely one of nature's means of dissipating energy. Thermodynamics and the Second Law are profoundly important in our attempts to understand change, but with regard the phenomenon of life, it is only part of the story. How do purpose, intentionality, cognition, and mind come into the picture? The physical basis that enables us to address such questions is the subject of the coming chapters.

7
Stability: The Role of Time

> In three words I can sum up everything I've learned about life: it goes on.
>
> —Robert Frost

D'Arcy Thompson, an insightful nineteenth-and-twentieth-century Scottish biologist and mathematician, once stated that "everything is what it is because it got that way." But *why* did everything get that way? The ubiquitous Second Law, described in detail in the previous chapter, does wonderfully in addressing change in the abiotic world, both here on Earth and in the wider cosmos, but as discussed, it is woefully inadequate when it comes to predicting change in the biological world. Biological evolutionary change is broadly directed toward increasingly *complex* and *unstable* material aggregates. Though Schrödinger explained that such behavior was not inconsistent with the omnipresent Second Law, it remains puzzlingly contrarian to the Second Law's underlying directive toward simplicity, stability, and equilibrium. And, of course, despite life's inherently unstable state, life has proliferated spectacularly and managed to occupy almost every conceivable ecological niche on this planet. Somehow the Second Law has become marginalized. Could it be that under certain circumstances, the ubiquitous Second Law ceases to be the dominant driver of observable change?

The answer to this question now seems to be *yes*. But when? Under what circumstances? Alfred Lotka, an insightful physical chemist, ahead of his time, noted a full century ago: "thermodynamics can only tell us what *cannot* happen. It does not tell us what *does* happen."[1] Nature, the ultimate technologist, found a way to have its cake and eat it. Let's see how.

We have learned that any physical or chemical system supplied with a continual source of energy can in principle circumvent the "downhill" Second Law directive, just as a car with gasoline in its tank can overcome the gravitational directive. But if a material system were freed from the Second Law's energetic constraints by tapping into some energy source, in what direction would change then take place? An organizing principle, which we have termed the *persistence principle*, begins to operate. Let me demonstrate how this principle is derived, how it can help explain change in the world at large, and how it can offer new insights into the life phenomenon. Remarkably, just like the Second Law, this general principle is exquisitely logical, but its controlling parameter turns out to be *time*, not *energy*.

Stability's Two Facets—Time and Energy

Let's look more closely at *stability*. The term was central in our discussion of the Second Law, where we focused on its energy facet. A material system is stable when its *energy* (more precisely, its free energy) has reached a minimal level, or equivalently, when its entropy has reached its maximal value.

However, stability has a second facet—*time*. In fact, that's the facet typically referred to in everyday language. If you are told that Tom and Jane have a stable marriage, you are not hearing about their energy or entropy. Their marriage is being categorized as long-term, or—as I put it—persistent. A time measure of stability is quite general. When we designate a weather pattern, a work environment, or a person's personality as stable, we are saying it is stable over time. The longer the time frame, the greater the level of stability.

Stability's time facet is relevant across the sciences. When we classify a particular radioactive isotope as being stable or unstable, we are not referring to the isotope's energy or entropy, but to its lifetime. For example, iodine-131, an isotope used for medical purposes, has a half-life of eight days, so it would be considered relatively unstable. Carbon-14, with a half-life of 5,700 years, is much more stable. A substance's half-life, expressed in units of time, would be a quantitative measure of its stability. Thus, stability can be about *energy/entropy*, or it can be about *time/persistence*. That fact, of itself, might seem puzzling. How can the same scientific concept mean two quite different things?

Even more confusingly, the two kinds of stability for a given material system can be contradictory, and that is awkward! Consider, for example, the standard reaction of hydrogen and oxygen gases ($H_2 + O_2$) to yield water (H_2O). Is such a gas mixture stable or unstable? It depends! An energy diagram for the reaction illustrates how the two types of stability play out (figure 7.1). The hydrogen-oxygen gas mixture is *unstable* in an energy sense, as its (free) energy is higher than that of the water product, ($\Delta G < 0$). Indeed, if you spark a mixture of those two gases, you will be rewarded with an explosion due to the heat that is generated by the reaction to form water. Nonetheless, even though that mixture

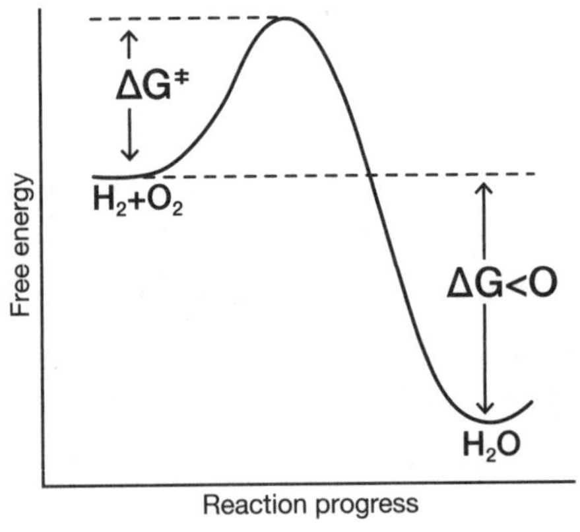

Figure 7.1
Free energy diagram illustrating the free energy change for the reaction of hydrogen and oxygen gases ($H_2 + O_2$) to give water H_2O ($\Delta G < 0$). Source: Author created.

is unstable in an energy sense (thermodynamically), it may be *stable/persistent* in a time sense (kinetically). If a spark or catalyst is *not* provided, then the gas mixture can remain unchanged almost indefinitely. An energy barrier separates the hydrogen and oxygen reactants from the water product ($\Delta G^{\ddagger} > 0$), so the reaction can only proceed if that barrier is overcome with a spark of energy. The study of reaction rates is the domain of chemical kinetics and is distinct from that of thermodynamics. Reactions can be fast or slow depending on *barrier heights* and can be independent of thermodynamic considerations.

These ideas are well understood in physics and chemistry. Indeed, as Lotka pointed out, both kinetics and thermodynamics determine what actually takes place in such processes; both factors need to be taken into account. Marginalize either at your peril. As we will discover in due course, much of the confusion associated with life issues in general, and the evolutionary process in particular, has arisen due to misunderstandings regarding the relative importance of kinetic and thermodynamic factors. Kinetic factors tend to govern life processes—*life is fundamentally a kinetic phenomenon, not a thermodynamic one.*

So the answer to the earlier question of whether a mixture of hydrogen and oxygen gases is stable or unstable is that it is certainly thermodynamically *unstable* (focus on energy), but it may well be kinetically *stable* (focus on time) depending on reaction conditions.

Broadly speaking *thermodynamic factors govern energy-stability, kinetic factors govern time-stability*. Of course, once a system has reached its equilibrium state, it cannot undergo further change, so, by definition, being energy-stable means it is also time-stable. Returning to our example of water, the water molecule (H_2O) is both energy-stable *and* time-stable. Water, suitably isolated, will remain unchanged indefinitely.

A *Venn diagram* can be useful for showing the logical connection between time-stable and energy-stable systems, as shown in figure 7.2. The category of energy-stable entities (represented by the shaded area) is just a *sub-category* of time-stable systems. Thus, energy-stable systems are necessarily time-stable (persistent), but time-stable systems are not necessarily energy-stable. An important conclusion then follows—time-stability (persistence) is more general than energy-stability. That in turn suggests that a law of change based on energy

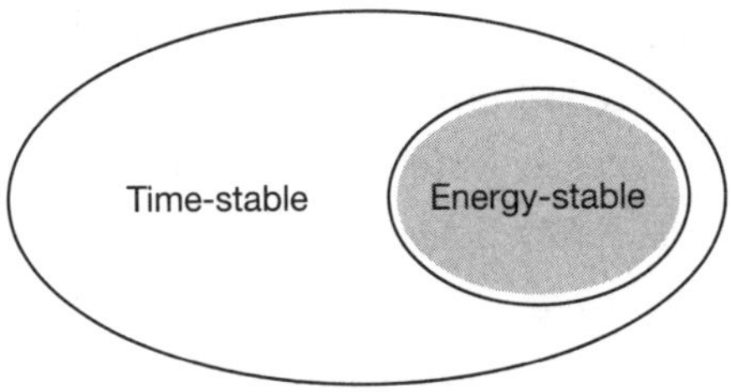

Figure 7.2
Venn diagram illustrating the relationship between time-stable and energy-stable entities within the general category of persistent systems. Source: Author created.

stability may well be *more limited* than one based on time stability.

That conclusion is quite remarkable. It suggests that there may well be a rule of change in nature based on *time* that is *more* general than one based on *energy*. Furthermore, such a law could be highly relevant to the life phenomenon. As discussed previously, life is a persistent material form, *unstable* thermodynamically, and so, by definition, belongs within the time-stable category, namely, in the unshaded segment of the Venn diagram. Let me now continue that line of thinking and seek the basis for what we have termed the persistence principle and explore how it can help explain central aspects of the life phenomenon.

Things That Persist

In a recent monograph entitled *Creation: Life and How to Make It*, Steve Grand, an insightful British computer scientist, made the following intriguing statement: *things that persist, persist, things that don't, don't*, claiming it to be the most important law of nature.[2] That's quite an ambitious claim and one that at first sight seems hard to justify. After all, the statement is a tautology, a statement that is necessarily true by virtue of its logical form. You might think that being a tautology, it cannot contain any new information about the world, and so it cannot be a law of nature—let alone the most important law of nature. Well, surprisingly, tautologies can be informative. Indeed, thinking more deeply about Grand's tautology, suggests that change *can* take place in the world, though it does not have to. Some things change, some don't, and the world around us mirrors that variable reality. That's quite informative for what appeared to be just a tautology.

Let me now reformulate Grand's statement so that it becomes more operative with explicit regard for the direction of change in the natural world:

Matter is driven toward increasingly persistent forms.

This revised formulation holds that there is a general direction for change—from *less* persistent to *more* persistent forms. If you think about it for a moment, the statement is logically based given the following: *Unchanging things don't change, and changing things do change, until they change into things that don't* . . . That statement explains the behavior of rocks and snowflakes quite well. Thus, the persistence principle in some sense states the obvious: the logical tendency of things to change until they stop changing.

Note that even if change toward more persistent forms is logically expected, that change does not have to be monotonic. There is no *a priori* reason for a *more* persistent entity not to become transformed into a *less* persistent one on occasion. Without further physical and chemical knowledge, we cannot comment on the direction of any *particular* act of change. However, if on occasion change does spontaneously occur in the opposite direction, from *more* persistent to *less* persistent, then the product of that transition will be relatively transient—by definition—and therefore susceptible to further change. Over time, therefore, broadly speaking, the direction of change will be, in ratchet-like fashion, directed toward more persistent forms.

But if this principle is no more than a logical statement, what then does it actually teach us, and how does it differ from the more famous general law of change in nature, the Second Law of Thermodynamics?

Kinetic Stability

For energy-stable systems (within the shaded area of figure 7.2), the persistence principle is not particularly useful. It merely provides in qualitative form information that the Second Law provides in quantitative form. However, the Venn diagram reveals something more—a second category of persistent forms—time-stable ones, which are *not* energy-stable. As noted, *kinetically stable* systems are time stable, so they occupy the white area of the Venn diagram.

But here's the thing: kinetic stability does not have to be of the static kind as exemplified by the hydrogen-oxygen gas mixture. Something can be persistent if it is being formed and removed at a rate that enables it to maintain a continuing presence. Stability can be of a *dynamic* kind. The impressive fountain in Geneva mentioned in the previous chapter is a physical example of such a system. The fountain is a time-stable physical entity even though the water that makes up the fountain is turning over continually.

Returning to chemistry, some years ago we predicted the likely existence of matter in a *chemical* dynamic state—matter that is being simultaneously formed and degraded chemically, resulting in a dynamic persistent steady-state, analogous to the fountain.[3] Though the physical scientist Willard Gibbs outlined a thermodynamic boundary that seemingly limited chemical possibility a century ago, that classification overlooked the possibility of chemical systems in a dynamic state. And indeed, two Dutch chemists, Jan van Esch and Rienk Eelkema, broke through that thermodynamic glass ceiling fifteen years ago, when they succeeded in preparing matter in that dynamic state, thereby confirming our theoretical prediction.[4] Effectively, a new material state of matter had been uncovered—a *dynamic kinetic state of matter*—and that has turned out to be significant. Research in material science is investing heavily in efforts in exploring the potential products of this new domain of dynamic kinetically stable (DKS) systems.[5]

But the discovery of persistent dynamic systems is significant for another reason—for its biological implications. Through the realization that dynamic kinetic states of matter exist, living things no longer need to be considered an inexplicable material form, seemingly inconsistent with established physicalist principles. The existence of DKS systems demonstrates that long-lived, energy-consuming, self-maintaining chemical entities are physically possible. In fact, the existence of such states open the door to a new chemical domain—that of *dynamic kinetic chemistry*. A detailed description of that domain, the basis for its existence, and its biological implications are the subject of the next chapter.

But before we move on, let me emphasize the key point of the previous discussion—the space of persistent chemical forms is not confined to matter that is thermodynamically stable. It also includes matter in a dynamic kinetically stable (DKS) state. Kinetic states of matter exist, with life being its most striking example.

8
Chemistry's Kinetic Dimension

> Nothing spoils a good story like the arrival of an eyewitness.
>
> —Mark Twain

In 1903, Wilbur and Orville Wright made the first powered flight. It only lasted fifty-nine seconds, but that landmark event opened the third spatial dimension to human travel. Something no less dramatic happened when a largely unrecognized domain in chemical potentiality—a kinetic domain—was uncovered. Though we are just at the beginning of this chemical voyage of discovery, it is already clear that a more fundamental understanding of life has come down to understanding the distinct chemical behavior and properties of matter found within that kinetic domain.

Initially, you might think that the claim of a new region in chemical space in a well-established science like chemistry might be somewhat exaggerated. But the clearest evidence for the existence of such a domain has been staring us in the face for well over a century—it is starkly evident in the behavior of all living things. A century's attempt at explaining life behavior in "regular" chemical terms has been woefully inadequate. True, the chemical activity associated with life's molecular bits and pieces all conform meticulously to established chemistry principles. Yet the holistic outcome of that frenetic activity makes little, if any, chemical sense. Think of life's unique attributes of agency, cognition, and mind. Think of ontogeny and morphogenesis and how strikingly different they are from "regular" chemistry. Think of a living cell and a dead one some seconds after it has died. Despite appearing almost identical, their subsequent behavior is as different as night from day. How can a living cell's "aliveness" be understood in chemical terms? As will now be described, understanding the life phenomenon begins with understanding the nature of the kinetic state.

Jan van Esch and Rienk Eelkema's startlingly simple modification of one of the most basic reactions in organic chemistry, esterification, demonstrated both the reality and the underlying potential of what we would classify as dynamic kinetic chemistry. When the esterification reaction is carried out in the traditional way, by mixing the carboxylate anion with a methylating agent, a methyl ester product is formed, often as a crystalline solid. Here's the equation.

$RCOO^-$	+	$CH_3\text{-}X$	$\rightarrow$	$RCOO\text{-}CH_3$	+	X^-
acid anion		methylating agent		methyl ester		by-product

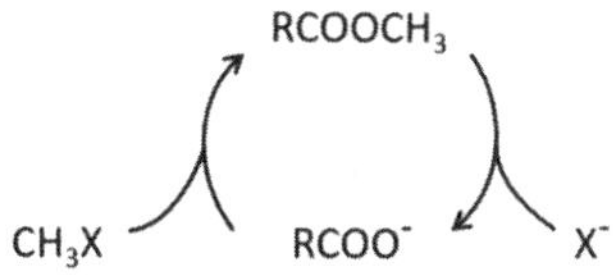

Figure 8.1
Schematic representation of a chemical process involving dynamic interconversion of carboxylate anion, $RCOO^-$, and methyl ester, $RCOOCH_3$, resulting in formation of a hydrogel in a chemical DKS state (composed of a dynamic aggregate of $RCOO^-$ and $RCOOCH_3$). Source: Author created.

However, van Esch and Eelkema discovered that if the reaction is carried out under *dynamic nonequilibrium conditions*, if resources are continually fed into the reaction mixture to generate the ester product while simultaneously *degrading* that ester product back to carboxylate anion reactant, as depicted in figure 8.1, a hydrogel, a type of soft material formed by the crosslinking of water-attracting polymers, is obtained. That hydrogel product, composed of a dynamic mixture of ester and carboxylate anion, is very different in its properties from traditional chemical products.[1] It does not exist in a traditional thermodynamic state, say, as a solid or a liquid, but rather in an unfamiliar material state, one we would term a *dynamic kinetically stable (DKS) state.*[2]

To better understand the nature of the DKS state, it is useful to think of that fountain image in the previous chapter. The fountain is a dynamic, nonequilibrium physical system that depends on a continual supply of water and energy to be maintained. The hydrogel product in the van Esch–Eelkema experiment is also in a dynamic state, just like the fountain. But in this case it is a *chemical* system, though both are maintained through the continual supply of material and energy. Metaphorically, one might think of that hydrogel as a *chemical fountain*—chemical matter in a dynamic, nonequilibrium steady state.

Chemically speaking, matter in that kind of dynamic, nonequilibrium state is unexpected. Stable objects in the material world—rocks, sand, crystals—are invariably in a static state. Their molecular constituents do not change with time. Van Esch and Eelkema had confirmed that matter could be generated and maintained in a persistent state despite being dynamic and energized. But here's the thing, and the central point of this text—all living things (excluding certain biological entities, such as spores and those in a cryptobiotic state) can be understood as being in that DKS state, and that realization offers a means of relating chemical and biological phenomena.[3]

DKS's Valuable Characteristics

When chemists prepare an organic ester, or any other chemical material in the normal (thermodynamic) way, one primary product is frequently obtained. Nothing unusual about that. But when the process is carried out in the *dynamic kinetic* manner, an *infinite* number of slightly different products can potentially be formed, depending on the specific reaction conditions employed. To help understand that plurality of products, think of chemical space as a multi-storied tower where the thermodynamic product resides on the ground floor, and all the other floors represent a set of possible kinetic products, with each member of the kinetic set having somewhat different properties. Accordingly, kinetic space is far more extensive than thermodynamic space, meaning that kinetic space is considerably *richer* in structural potential compared to thermodynamic space. For example, when preparing hydrogels in the DKS state, a stiffer, stronger, and more elastic hydrogel can be generated under one set of kinetic conditions, while under slightly different conditions, a softer, more flexible hydrogel might form. In fact, the hydrogel's properties can be *fine-tuned* to take on a variety of desired physical characteristics. That is significant since such structural flexibility is not accessible within the traditional thermodynamic regime. It explains why the discovery of intelligent soft materials systems mimicking the dynamics and behavior of living systems has opened the door to new possibilities in material science (for a review, see note 5 in chapter 7).

A particularly important characteristic of DKS systems is that they are *adaptable*. Think of that physical fountain analogy for a moment. A sudden gust of wind can cause the fountain to spontaneously tilt to one side, though the fountain's dynamic state continues to be maintained. And once the wind gust subsides, the fountain reverts to its original form. That simple physical analogy illustrates the inherent adaptive and "self-healing" capabilities of DKS systems. DKS systems respond instantly to changes in their environment by shifting into an adjacent DKS state more suited to the new conditions. Significantly, being able to shift between different DKS states depending on contingent circumstances, enhances the DKS state's *global* stability. Indeed, such dynamic versatility is central to life's extraordinary resilience.

Thus, one might generalize by saying that are two distinct regions of chemical possibility within chemical space—the traditional thermodynamic domain that chemists have been exploring for close to two hundred years, and the recently discovered dynamic kinetic domain, as shown schematically in figure 8.2.

But now to the key point of this discussion: as illustrated in figure 8.2, all living things—bacteria, giraffes, elephants, and you reading this text—lie within that dynamic kinetic domain. Broadly speaking, most life's processes taking place in you at this moment are *not* traditional ground-state chemical processes of the kind taught in undergraduate chemistry courses but rather *energized dynamic processes*. Living stuff is fundamentally matter in the process of continually being turned over, and it is the flexibility associated with matter in that dynamic state that enables living organisms to carry out life's distinct and very special processes. Though systems chemists formally discovered that kinetic domain

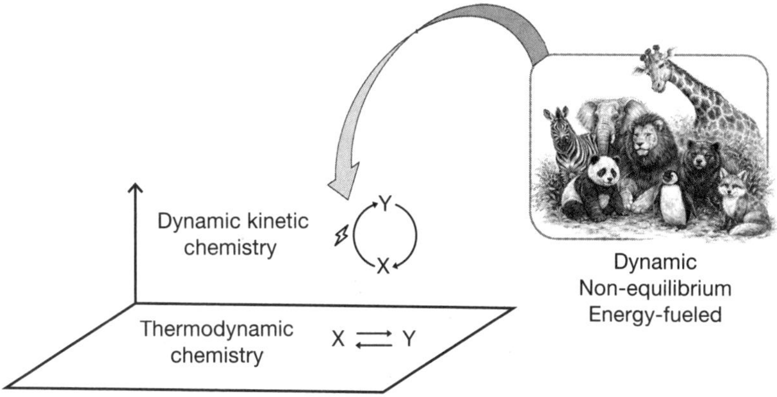

Figure 8.2
The existence of two reactivity domains in chemical space, schematically represented by X and Y interconversion—one in the traditional thermodynamic domain, and one in the recently discovered dynamic kinetic domain. As illustrated, all living things lie within the dynamic kinetic domain. Source: Author created.

quite recently, nature has been successfully exploiting dynamic kinetic chemistry for billions of years. Let me now consider how DKS systems could have emerged naturally.

Kinetic Power of Replication

The existence of any DKS state depends on matter in that state undergoing continual formation and degradation processes, thereby generating that "chemical fountain". A reaction optimally able to establish such a dynamic process is the *replication reaction* due to its extraordinary kinetic power. As Thomas Malthus already appreciated over two centuries ago, replication, being exponentially driven, is unsustainable. Thus, a persistent replicative state could be established and maintained if a counter-balancing degradation process, converting the replicator back to its component building blocks, were to take place simultaneously. Furthermore, once the system is in that replicative steady state, the stability of the state would not be thermodynamic, but rather dynamic kinetic. Its stability would manifest in the steady-state *population* of replicators, not in the individual replicators, which are transient, like water drops in a fountain.[4]

You might think that replicative DKS states, being in that dynamic steady state, might be fragile, relatively unstable, of low persistence. But that's not necessarily the case. Replicative DKS states can be more stable than even the hardiest of passive materials. Consider cyanobacteria, an ancient life form. Like all life forms, cyanobacteria are in that DKS state, yet they have managed to exist continuously for over 3.5 billion years. That *is* stable! In fact, it is only through the extraordinary stability that replicative DKS systems can attain that life is possible.

The critical role of the replication reaction in the possible establishment of a DKS state was vividly illustrated in a computational study by Blaise Agüera y Arcas, an American software engineer at Google Research in 2024.[5] He investigated random variation in instruction-like sequences within a "primordial soup" of coding segments, and his study revealed that given enough time, coding segments would eventually stumble upon self-copying sequences, after which exponential dynamics took over. Random exploration through sequence space had eventually led to a sequence with a replicative capability. Reproduction had spontaneously emerged from a nonreplicative environment—a striking finding!

But there was more. Once reproduction had begun, the initial randomness in the system's dynamics quickly became ordered and the subsequent evolutionary dynamics changed dramatically, reflecting that replicative capability. The kinetic power of replication and its likely consequences had been clearly demonstrated, though, of course, such a computational study would not be expected to mirror behavior in the physical universe in a precise way. Computational dynamics alone do not take into account physical constraints that would necessarily apply in a physical system.

Life as Dynamic Kinetic Chemistry

Let us now explore how life has utilized the DKS state to uncover and exploit life's most extraordinary capabilities, including consciousness and mind. Nature discovered that within the replicative DKS world, the sky is the limit. Let's see how.

Consider life's "aliveness," that extraordinary yet elusive property that distinguishes living from nonliving forms. Living systems aren't just molecular clumps of stuff. They are dynamic assemblies in continual interaction with their environment, systems that need to be continually fed resources to be maintained. Well, chemical DKS systems, such as van Esch's hydrogel, behave in just that way. They do the equivalent of "eat" (utilizing supplied chemical resources), "excrete" (by eliminating reaction side products), and dissipate energy, just like elephants and bacteria. What we characterize as "aliveness" in a biological system can already be discerned in rudimentary form in chemical DKS systems.

What then is death in chemical terms? Death is just the collapse of a DKS state to its underlying thermodynamic state. Once the supply of energy or material resources is severely impacted, or ceases, for whatever reason, the DKS state collapses—it "dies." And getting the thermodynamic state back into the energized DKS state is no trivial matter. Several factors need to be operational simultaneously for the dynamic state to be regenerated. As systems chemists know only too well, chemical DKS states are not readily prepared! No surprise then that attempts to mimic life's emergence by mixing and reacting life's molecular components in a traditional chemical manner were unsuccessful.

Let us now consider some examples of how the inherent flexibility of DKS states provides life processes with their exceptional functionality.

Just as you and I possess a skeleton, the biological cell also has an analogous structure, the cytoskeleton. It is composed of a dynamic network of protein filaments and tubules in the DKS state. That chemical network not only provides the cell with structural support, but also enables the cell to fulfill a variety of other functions. Through its dynamic capability it can initiate cell motion and facilitate material transport within the cell. It also participates in cell division and in the relaying of chemical signals into the cell's interior. All these functional capabilities are enabled through DKS-state fine-tuning of the cytoskeleton. The cytoskeleton DKS system is stunningly efficient. It is one further manifestation of the cell's "aliveness." But that's just the beginning.

Dynamic control within the cell pervades all biological processes, and the cell's metabolic processes, invariably cyclical, are central to that dynamic control. Given the fundamental role proteins play in cellular function, protein regulation is critical, and a process termed "ubiquitination" has evolved to regulate the fate, location, and activity of proteins.[6] Thus, intracellular proteins are continually degraded and renewed through dynamic turnover to ensure protein fidelity. That process is initiated by proteins designated for degradation being labelled with a small protein, ubiquitin. Once identified by the ubiquitin label, the protein is broken down by cellular action into its amino acid constituents.

The same dynamic behavior found in protein metabolism is also observed in other cellular structures. The cell membrane is thus not a static structure like an egg shell, but a dynamically functional entity, active in regulating material flow into and out of the cell. If dynamic turnover of the cell's membrane molecules were to cease, the membrane would lose its functionality; it would break down and release the cell's contents.

Life's dynamic character also manifests at the cell level. The cell acts as a dynamic, energy-driven, nonequilibrium system, with most cells in your body being continually turned over. In fact, there is very little of you now reading these pages that will still materially be you some months from now. Molecule-by-molecule, cell-by-cell, you are continually turning over. Homeostasis, the biological hallmark of life, the process by which cells maintain their energized nonequilibrium dynamic state, is just the biological term that describes the maintenance of the DKS state at the organelle, cellular and organismic levels.

Finally, many of us might presume that at least some structures in the human body are static—for example, our mammalian bones, like the rafters and girders in the structure of our dwellings. Well, that's not the case either. Bones are also dynamic systems. Bones undergo continual formation and resorption processes, and it is that dynamic character that allows young bones to grow and broken bones to heal. To facilitate that dynamic process, specialized bone cells induce bone formation, while a second cell type is responsible for bone resorption, the breaking down of bone tissue. Yet again, another example of life's dynamic nature. All living things are ultimately constructed from life's central building blocks—*matter in the DKS state.*

So What Is Life?

In simplest terms, life could be thought of as a *chemical replicative network in the DKS state.* That description captures the core physical/chemical claim this book advances. Life's existence and functionality depend on the two factors specified: being replicative and in the DKS state. It is the replication process that enables the emergence of a Darwinian evolutionary process. But being in the DKS state is also an essential requirement. Close to seventy years of experience with "regular" chemical replicative systems has taught us that replication alone is not sufficient to generate a life-like evolutionary process. Recall Spiegelman's classic RNA molecular evolution experiment in which the evolutionary direction was "downhill," directed *away* from life, not *toward* life. Replicative chemical systems exist, but if they are not in the DKS state, they do not evolve toward life. The inability of RNA molecules to evolve in a life-like evolutionary process was the main reason the RNA-world theory ran into conceptual difficulties.

In contrast, a replicative system in the DKS state—being dynamic, energized, nonequilibrium, even cognitive to a degree—is in some rudimentary sense already exhibiting signs of "aliveness". It already mimics the replicative dynamics found in living systems. As described by Weizmann Institute scientist Shneior Lifson, the math describing the kinetics of such systems indicates that two DKS replicators that compete for a common resource cannot coexist.[7] The mixture is predicted to evolve towards the extinction of the less efficient replicator, thereby enabling classical evolutionary behavior. Indeed, recent experiments from systems chemist Sijbren Otto's University of Groningen laboratory on competing DKS systems lend support to those theoretical predictions.[8] Thus, in the replicative DKS world, evolutionary change from *less DK stable* to *more DK stable* is expected. That direction of change contrasts with the direction of change in the "regular" chemical world, from *less thermodynamically stable* to *more thermodynamically stable*. To summarize:

Direction of change in the replicative DKS world:

DKS *less* stable → DKS *more* stable

Direction of change in the "regular" chemical world:

Thermodynamically *less* stable → Thermodynamically *more* stable

This analysis suggests that for a chemical system to undertake a life-like evolutionary process in the laboratory, it would need (a) to be replicative, (b) to be in the DKS state, and (c) to be amenable to structural modification through imperfect replication. Alas, the experimental challenge posed in satisfying these three requirements remains a substantial challenge for systems chemists.

Evolution's Driving Force?

Stephen Jay Gould firmly held that evolution is fundamentally *nondirectional*, as it is contingently dependent on random events. That position matches the modern evolutionary view as expressed by the popular Oxford biologist Richard Dawkins in his book *The Blind Watchmaker*, as noted above.[9] The philosopher Daniel Dennett, together with other evolutionary biologists, have long supported that nondirectional and contingent view. The basic argument is that natural selection is an algorithmic procedure, that it is not forward looking, and therefore can have no direction in mind.

Yet murmurs of dissent persist. Simon Conway Morris, the Cambridge University paleontologist, has argued that evolution does have direction, though not in the traditional sense of having a precisely predetermined path. His work on evolutionary convergence, showing that different species evolve similar traits independently, suggested that evolution does have a degree of predictability, implying directionality of some kind.[10] Thus, sharks and dolphins, though not closely related, have evolved similarly streamlined bodies independently. Even more strikingly, the eye has evolved in functionally distinct ways in several lineages—camera-like eyes in mammals and cephalopods (squid and octopuses), compound eyes in arthropods (insects), and multiple eyes in spiders. The reality of convergent evolution clearly hints at some kind of directionality in evolution.

Though the evolutionary process is traditionally treated as a biological one, in academic circles at least, the process is presumed to have begun in chemistry. Life did not just emerge spontaneously out of nowhere.[11] That suggests the evolutionary process would have started in chemistry, though that fact tends to be marginalized. Let's dig into this conundrum.

Stochastic events, meaning events that are random, by definition lack a deterministic driving force or inherent direction. Brownian motion, the random movement of small particles in a medium due to collisions with other fast-moving particles, is a well-known example. Think of motes of dust in a sunbeam. Surprisingly, however, motion that is *microscopically* random can induce *macroscopic* directionality, and, strikingly, it does so in both the physical *and* biological worlds. Let's discuss each in turn.

Consider the ink drop diffusion experiment described in chapter 6. Even though the motion of each individual ink particle in solution is random, there *is* a macroscopic direction of change for the ink drop in its entirety, and it is anything but random. Macroscopic change in this case is clearly directed towards the even dispersion of ink throughout the medium. It is that omnipresent Second Law at work with its statistical underpinnings that drives the system toward the uniform equilibrium state. Bottom line: *macroscopic change in the nonliving world is directed, proceeds irreversibly, and has an identifiable driving force.*

But what about the world of dynamic replicators? It turns out that the same general principle applies in that world as well. If some random genetic mutation takes place within a replicating population, the DKS population (whether chemical or biological) would also be directed towards greater stability, but in this case toward greater *DKS* stability, as described earlier, thereby mirroring the Second Law directive. The math is different, but the consequence is the same. *In both cases, random change leads to nonrandom consequences.*

Accordingly, *all* macroscopic change in the material world is directed—toward greater stability. Within the "regular" physical/chemical world, the drive is toward greater *thermodynamic* stability, while in the living world, the drive is toward greater *kinetic* stability, as depicted in figure 8.3. Each path is governed by its particular mathematical directive: the math of probabilities directs the thermodynamic path, while the math of exponential growth directs the kinetic path. Thus, despite the evolutionary process being "uphill" energetically, and generally reflected in greater complexity, the evolutionary direction is unequivocal. It is directed toward greater stability, but stability of the DKS kind. Accordingly, evolution can be seen as having *both a driving force and a direction.* A common biological view, that evolution has no direction, appears to be incompatible with fundamental physical/chemical thinking: all macroscopic physical/chemical change in nature is directed.

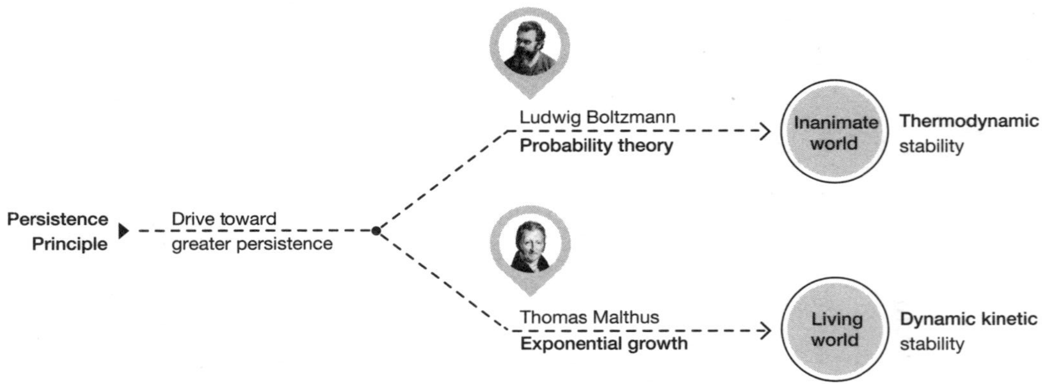

Figure 8.3
Diagram illustrating the two possible paths leading to greater persistence: the thermodynamic stability path, based on the math of probability theory, credited to Ludwig Boltzmann, and the dynamic kinetic stability path, based on the math of exponential growth and ideas of Thomas Malthus. Source: Author created.

An intriguing question now arises: why is the direction of change unambiguous in the thermodynamic world, but mysteriously indeterminate in the replicative DKS world? Inspection of the branching patterns for change in the two worlds offers an explanation. In "regular" chemical space, all chemical processes are directed toward their thermodynamic sink, like bathwater flowing into a drain, or a ball rolling to the bottom of a hill. The topology of thermodynamic space is *convergent*, as depicted in figure 8.4a. Consider the combustion of different hydrocarbons—say methane, propane, and butane. All yield the same stable products, carbon dioxide and water. That is the reason change in the "regular" chemical world is predictable. Statistical logic underpins the direction of chemical reactions in thermodynamic space.

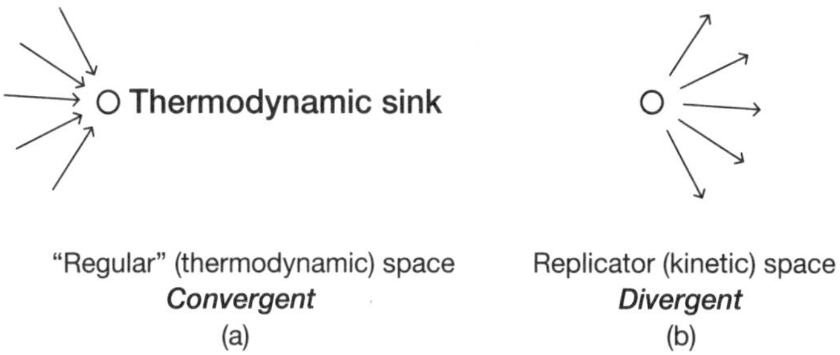

Figure 8.4
Schematic representation of branching patterns within (a) thermodynamic space (convergent); and within (b) dynamic kinetic space (divergent). Source: Author created.

In contrast, the corresponding branching pattern for change in the replicative DKS world is *divergent,* as depicted in figure 8.4b. For a replicative DKS system, the path towards increasingly persistent forms is *contingent* and dependent on specific conditions at that moment. Accordingly, DKS selection introduces direction without predetermination because there are multiple potential paths by which a replicating system could enhance its dynamic kinetic stability. The idiom that "there is more than one way to skin a cat" captures that divergent and weakly predictive nature of change in the DKS world. Indeed, there are some forty-one recognized cat species (family Felidae), grouped into two major subfamilies, Pantherinae (big cats like lions, tigers, and leopards) and Felinae (small to medium-sized cats like domestic cats, servals, and lynxes), reflecting evolution's divergent character. Note, however, that in the myriad ways of "skinning the DKS cat," significantly, none lead to the emergence of mushrooms or elephants. Thus, some degree of evolutionary prediction *is* possible, though far less definitively than in the thermodynamically directed world.

This thinking fits in with Stuart Kauffman's concept of the "adjacent possible," a powerful idea in complexity science and evolutionary biology.[12] The concept describes how new evolutionary forms, functions, and ideas do not emerge arbitrarily, but by exploring "next step" evolutionary possibilities, which are just one step away from the existing state. Life's limited evolutionary predictability derives directly from the divergent nature of dynamic kinetic space. The lack of appropriate physical/chemical grounding for the evolutionary process may have contributed to the confusion surrounding attempts over recent decades to formulate a coherent theory of evolution.

To summarize: every DKS system is in continual dynamic interaction with its environment, so its evolutionary path cannot be mathematically specified in the same way as for thermodynamically controlled processes. If you release a ball at the top of a bumpy slope, you may not be able to predict the exact pathway the ball will take on its downhill path, nor the ball's final location, but you can confidently predict that the ball will end up at a level lower than the one from which it started. The logic of evolution is unambiguous—the broad DKS prediction is that life evolves to become *more stable, more complex, more diverse.*

The Role of Complexity

Amid the vast diversity of stable life forms, highly complex organisms clearly predominate. As discussed in chapter 5, nature the technologist "discovered" from the outset of the evolutionary process that improved function is facilitated by greater complexity. Yet this immediately raises a fundamental question: can this directive toward improved functionality be physicalized?

Function, though foundational in biology, does not comfortably reside within the epistemology of physics. Matter and its constituent components—atoms, electrons, neutrons, quarks, and so on—have no function. They just are. So how did function in its teleological (purposeful) sense emerge from what would have "just" been a physical system? As will now be discussed, the DKS concept offers a viable framework for addressing this long-standing puzzle.

We begin with an important observation: the enormous diversity of functions exhibited by living systems today was not present at life's origins. They emerged gradually over time. Moreover, it is reasonable to suppose that this vast repertoire of biological functions arose from a much more limited set of early functional capabilities—perhaps even from a single primary function. I propose that this primary function was replication. Without replication, the living state could not have come into existence at all. It is this centrality that elevates replication to the status of life's foundational function.

The experimental difficulties encountered in creating simple molecular replicators underscore just how nontrivial this function is. Even relatively simple replicating entities, such as RNA molecules, do not replicate readily—a persistent frustration for systems chemists. Without precisely controlled conditions and material support, replicating molecules often fail to replicate. In living systems, however, replication takes on a radically different character. It is no longer fragile and conditional; it is robust, pervasive, and irrepressible. Bacteria, for example, replicate successfully in environments that would be utterly hostile to most life forms: high-radiation zones such as nuclear reactor cooling systems, the crushing pressures and near-freezing temperatures of the deep ocean, and countless other extremes. No laboratory, no experimentalist, no supervisory chemist on hand!

This extraordinary capability is not the result of chemical simplicity, but of extraordinary organizational complexity—structural and network—complexity refined and extended over billions of years of evolution. Indeed, it is life's total dependence on reliable replication that conferred upon replication its functional status. That functional status, in turn, establishes a causal relationship between *replicator stability* and *replicator complexity*. In the DKS framework, stability is not a passive property, but an outcome actively maintained through dynamic processes. Figure 8.5 illustrates this relationship: replicator function, replicator complexity, and replicator stability form a mutually reinforcing triad. Replicator function improves with replicator complexity, leading to enhanced replicator stability. Indeed, within life's branching evolutionary topology, there is one structural feature that can be predicted with some confidence: the long-term trend toward increasing complexity—greater complexity for enhanced stability. In a dynamic kinetic world, complexity acts as a buffer against catastrophe.

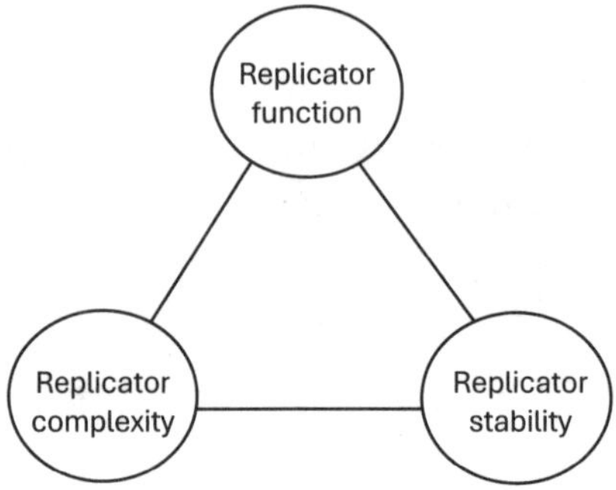

Figure 8.5
Diagram illustrating the causal connection between replicator function, replicator complexity, and replicator stability in the replicative DKS world. Source: Author created.

At this point it is worth repeating a previous comment: complexity in biology is not just structural, i.e., associated with the molecular form of individual replicators. *Network complexity* also plays an inordinate role; living systems do not exist or evolve in isolation. Animals, plants, bacteria, and viruses interact, compete, cooperate, and coevolve within richly interconnected networks. Over geological time, this web of interactions has grown to extraordinary levels of complexity—and corresponding stability. It was precisely that insight that motivated James Lovelock's Gaia hypothesis in the 1970s—the proposal that life on Earth and its nonliving environment together form a self-regulating planetary-scale network.[13] Though initially met with skepticism, the Gaia perspective has gained increasing empirical and theoretical support.[14] From this vantage point, life emerges as a fundamentally different kind of chemistry, one dominated not by individual reactions but by persistent, evolving *reaction networks*. In DKS terms, life represents a chemical domain governed by dynamic persistence rather than static stability. And the underlying directive of this domain is unmistakable: the drive toward increasing persistence through extended network organization. Indeed, the argument that is frequently made, that bacteria manifest life in its simplest form, tends to overlook the fact that bacterial populations can involve many millions of interacting bacteria. Bacteria are better understood as cellular networks rather than as a collection of individuals.

Which brings us naturally to the biological concept of cooperation, a term that captures the functional dimension of complexity. Cooperation is ubiquitous in living systems. As the evolutionary biologist Richard Michod has argued, "cooperation is now seen as the primary creative force behind ever greater levels of complexity and organization in all of biology." Similarly, complex-systems biologist Peter Corning introduced the concept of synergy to explain how cooperative interactions generate functional outcomes that no component could achieve alone.[15]

Taking a physical perspective reveals the underlying basis of these phenomena. What biology describes as cooperation, physics recognizes as dynamic kinetic complexification. The two languages emphasize different aspects of the same underlying process. Biology highlights the functional benefits; physics highlights the structural and kinetic prerequisites. Both lead to the same conclusion captured succinctly in the familiar phrase: *alone we can do so little; together we can do so much.*

The physical perspective, however, extends this insight a step further. It reveals that complexity is not merely advantageous—it is tightly bound to persistence itself. In the DKS framework, increasing complexity is not merely an evolutionary benefit; it is a physical requirement for long-term survival in a dynamic and unpredictable world.

We have unavoidably arrived at the question of whether life is likely to have emerged elsewhere in the universe. Though discussion on that issue is beyond the scope of this book, I will just say that by reducing life and its emergence to established chemical concepts, effectively to another kind of chemistry, the feasibility of life elsewhere in the cosmos is more readily assessed. With life now understood to be a means by which nature seeks persistent forms, that insight alone suggests that life should not be considered an extraordinary event. Although the role of contingent factors cannot be ignored, life increasingly appears to be an inevitable consequence of a cold, objective numbers game. It is out there fundamentally because, as Steve Grand says, things that persist, persist; things that don't, don't. Nature seeks persistent forms, and there exist two separate and distinct physical pathways towards persistence—(a) the *thermodynamic path,* reflected in "regular" physical and chemical processes, which leads to increasingly persistent forms through energy dissipation and the drive toward simplicity and disorder, and (b) the *dynamic kinetic path*, reflected in life processes, which lead to greater persistence through increasingly complex, cooperative replicative networks. Both processes rest on established mathematical formalisms. Both processes express the inherent logic of the persistence principle. Both processes are expected to manifest throughout the universe, though life's contingent nature makes it less clear how prevalent dynamic persistent material forms are likely to be.

In the final two chapters of this text, we will discuss two of the most challenging life questions: the physical basis for purpose and the physical origin of mind—how these subjective characteristics, seemingly incompatible with a physicalist world, were able to emerge from an objective universe.

9
Physicalizing Purpose

> There is nothing more deceptive than an obvious fact.
>
> —Arthur Conan Doyle

No issue in biology has caused more angst and confusion than that of agency, that sense of purpose that appears to govern the structure and activity of all living things. Organisms appear to act with intent, as if pursuing clear, goal-directed outcomes. Watch a lion in hot pursuit of a deer in a frenzied dash to escape and there is no doubt as to the purpose of both lion and deer. But there is no need to go on safari to see purpose on display. It's there all around us, all the time, both in life's structures and in its behavior. Just as you don't have to be a Newton or an Einstein to acknowledge the reality of falling apples, you don't have to be a Wittgenstein to acknowledge the existence of purpose. And yet, for over a century, modern biology has struggled with how to characterize it—if at all. Many in the field have sought to marginalize or outright reject its legitimacy as a biological concept.

So what exactly is the problem? Though Aristotle formalized teleological thinking with the idea that natural phenomena have inherent purposes or goals (*telos*) toward which they strive, since Darwin's landmark theory of evolution, many biologists have felt uncomfortable with such thinking. The scientific issue that arises is fundamental. The seventeenth-century scientific revolution, brought about by figures such as Copernicus, Galileo, and Newton, ushered in a view of nature as objective, mechanical, and devoid of purpose, a view that remains deeply engrained within physics. Mention the word "purpose" to a physicist and he will likely grimace in discomfort. The word does not belong within the lexicon of physics. Physics is concerned with quantifying nature's fundamental laws, and purpose plays no role in the formulation of those laws. Jacques Monod famously captured the essence of that scientific revolution in just three words: *nature is objective*. In fact, the importance of Darwin's theory of evolution stemmed from the fact that in its formulation, Darwin had taken the first major step toward the physicalization of biology. Natural selection, Darwin's conceptual linchpin, was a logical, objective process. Purpose was not involved. Post-Darwin, biological thinking sought to fit in with that "nature is objective" perspective.

Yet, a fundamental difficulty immediately asserts itself, one that is impossible to ignore. If nature is objective, as science claims, how could insensate matter—atoms and molecules, the stuff of rocks, water, and air—have come to life? How could certain material systems, those we classify as living, express purpose, intentionality, goals, and desires? As Jacques Monod put it pointedly: here lies "a profound epistemological contradiction."

Awkwardly, the more the wonders of life were revealed through molecular biology's mechanistic and structural studies, the deeper the mystery became. Life's apparent purposiveness undermines biology's central tenet that biological systems are ultimately nothing more than physically directed entities. To understand how living and nonliving worlds relate to one another requires resolution of that longstanding paradox.

Is Purpose Real?

Let me begin by noting that within traditional biological thinking, there is one school of thought that outright rejects the concept of purpose (teleology) as an explanatory framework for biology. Building on classical Darwinian thinking, it is argued that the appearance of purpose is an illusion that arises through organisms following instructions coded in the organism's genome. Adaptive traits that arose through Darwinian natural selection, became encoded in genes making them *seem* purposeful.

But that view has become increasingly untenable. As pointed out by Scott Turner, the State University of New

York biologist, purposeful behavior is too obvious to be casually dismissed in this way.[1] For example, homeostasis, the process by which organisms regulate internal conditions (temperature, blood pH, and ion concentrations, among others) is explicitly purposeful in its operation. The homeostatic system does not just respond mechanically to perturbations—it also anticipates and compensates for them. Homeostasis is a multi-layered, adaptive system that integrates responses at the molecular, cellular, organismic, even behavioral level, to ensure life operates within allowable limits, in an environment that is inherently unpredictable.

With time, additional expressions of purpose continue to be discovered deepening the dilemma. Immune cells, for example, avoid obstacles when moving toward targets. They adjust their paths and reach specific locations through a problem-solving approach that extends well beyond genomic directedness. Animal regeneration is another striking example. Michael Levin, a Tufts University biologist, recently demonstrated that if planarian flatworms are cut into pieces, each piece regrows into a complete organism. The process is purposeful in that there is a nongenetic, system-level memory of "what the organism should be." Levin discovered that the goal of regeneration is encoded in nongenetic patterns of electrical connectivity, much like memory in a neural network. Such a stable, reprogrammable, system-level goal acts as a morphological "set point," guiding cells during regeneration. Behavior aimed at re-establishing specific anatomical goals is reasonably deemed purposeful.[2]

But there's more. We have recently learned that purposeful behavior is not just characteristic of complex life forms but is already apparent at the simplest life level—in bacteria. Bacteria routinely express complex decision-making capabilities revealing highly purposeful activity. It led James Broach, a Princeton biologist, to conclude in a 2007 paper that bacteria can "think," and some years later Pamela Lyon, a microbiologist at the University of Adelaide, made the far-reaching claim that bacteria do the equivalent of "talk, chat, gossip, cheat, vote"—purposeful activities for sure, all in pursuit of their bacterial aims.[3] Bacteria are now widely seen as social creatures that chemically communicate with one another through a process termed "quorum sensing."

In part due to the growing numbers of studies akin to those mentioned above, the Third Way of Evolution, a movement within evolutionary biology launched some years ago by James Shapiro, Denis Noble, Raju Pookottil, and others, has firmly endorsed the purpose concept. In fact, in a recent monograph entitled *Evolution "On Purpose,"* a group of leading biologists took the argument a bold step further. Beyond claiming that individual life forms act purposefully, they presented evidence that indicated that the evolutionary process itself is purposeful, that the evolutionary direction is intended.[4]

For traditional biology such a claim is close to sacrilegious, striking at the very heart of established evolutionary theory. But newly discovered facts cannot be ignored. One strong piece of evidence supporting the reality of purpose in evolution was obtained from the study of the bacterial immune system. Bacteria can defend themselves against viral attack through a natural bacterial gene-editing technology named CRISPR-Cas9, whose discovery earned Jennifer Doudna and Emmanuelle Charpentier the 2020 Nobel Prize in Chemistry.[5] The mechanism involves the bacterium capturing snippets of viral DNA present from earlier viral attacks and incorporating them into its genome. That genomic change assists the bacterium in identifying an impending viral attack enabling it to "kill" the attacking virus by cutting its genomic strand into two. The point is that *organisms purposefully modify their own genome as part of their struggle to survive*. Nothing random about that. As James Shapiro pointed out: "The capacity of living organisms to alter their own heredity is undeniable." That idea counters a century of traditional evolutionary thinking built on the central pillar of random genetic variation.

The academic consequences are far reaching. A rethink of one of biology's central assumptions, that evolution has no direction, is required. Furthermore, for a revised theory of evolution to be comprehensive, it will also need to integrate life's *nonphysical* aspects, in particular mind. While the extended evolutionary synthesis (EES) did resolve some deficiencies of the modern synthesis, biology's nonphysical aspects have remained unexplained, mysterious as ever.

Looking back, what went wrong? How could physicalist thinking, by denying the role of purpose and intentionality in life, have undermined biological theory for almost a century? Well, actually, it didn't. The leading physicists of the twentieth century—Niels Bohr, Erwin Schrödinger, and Eugene Wigner—almost a century ago openly stated that biology remained physically inexplicable. As Erwin Schrödinger put it: "Living matter, while not eluding the 'laws of physics' as established up to date, is likely to involve 'other laws' of physics hitherto unknown."[6] Niels Bohr went further: "The existence of life must be considered as an elementary fact that cannot be derived from physical and chemical laws."[7] And Eugene Wigner expressed a similar sentiment: "The full meaning of life, the collective meaning of all human desires, is fundamentally a mystery beyond our grasp."[8]

What these leading physicists were effectively saying was that the "nature is objective" axiom, while accurately reflecting the nature of change in the nonliving world, was clearly inapplicable in the living world. No matter, the physicalist view of life continued to dominate biological discourse, and consequently the contradiction between physicalist thinking and biological reality continued to impede progress toward a biologically coherent theory of life. The central challenge for the natural sciences remained unchanged: to formulate a theory of life and evolution able to integrate life's *non-physical* aspects, including those of purpose and mind, in a physically acceptable way. Let's see how reaching back to the beginnings of the evolutionary process might assist in that endeavor.

Persistence Meets Purpose

So how can purpose be understood in physical terms? In a nutshell, the biological concept of purpose can be linked to the physical concept of *persistence*. Let's see how.

As discussed earlier, kinetic states of matter exist and maintain their persistence through dynamic self-manufacture and self-maintenance. Indeed, through such activity they take on autonomous character. They adapt to perturbations and seek to maintain their persistent state, and that behavior would be expected to apply to all DKS systems, whether chemical or biological. Thus, the biological phrase "seeking to stay alive" could be expressed in physical terms as "seeking to persist." The following passage from Michael Crichton's science fiction novel *Congo* helps make the connection.

> "The purpose of life," Munro said, "is to stay alive. Watch any animal in nature—all it tries to do is stay alive. It doesn't care about beliefs or philosophy. Whenever any animal's behavior puts it out of touch with the realities of its existence, it becomes extinct . . ."
>
> "Maybe there is a higher truth than merely staying alive," Ross said. "There isn't," Munro said.

That's it. Purpose, as an emergent biological reality, exists, and life's central purpose, expressed in unadorned physical terms, is to persist. Thus the common-sense drive to stay alive finds expression within the persistence principle, the simple idea that the direction of change in the natural world takes place from less persistent to more persistent forms. The principle applies in both the "regular" physical/chemical world, where it is governed primarily by thermodynamic factors, but also in the kinetic world of DKS replicators, where kinetic factors are dominant. In the "regular" physical/chemical world, the persistent principle can be expressed quantitatively in entropy/energy terms. In the biological world, as discussed in chapter 8, quantification can only be undertaken in limited cases, such as for replicating DKS systems competing for common resources, and then only in relative terms. In most cases, say, comparing the persistence of bacteria to that of elephants or mushrooms, only qualitative statements can be made. Thus, bacteria, having existed for several billions of years, would be considered more persistent than elephants (members of the Elephantidae family) that have only existed for a mere six million years, and even more persistent than mushrooms (broadly, fungi), that have existed for some 700 or 800 million years.

But having said that, let's be quite clear: concepts in science that are largely qualitative can be informative, even highly so. Darwin's principle of natural selection was entirely qualitative, yet revolutionary in its impact.

The concept of purpose offers insight into multiple biological phenomena, and its wide utility explains why biological thinking has been unable to suppress its usage despite decades of seeking to do so. Life's central purpose is to persist, though the means of persistence are varied and not readily quantifiable. But let's keep in mind the essence of the scientific method: in the broadest sense, science is ultimately an exercise in pattern recognition, and in biology, where quantitative relationships are not always feasible, the concept of purpose certainly offers insight. As William Bruce Cameron, an American sociologist, noted: "Not everything that counts can be counted, not everything that can be counted, counts . . .". Purpose, as a concept, certainly counts!

Life's Primary Function

Our discussion on purpose necessarily leads us to the concept of function, as the two terms are closely related. In biology, function refers to the role or activity that a structure or process performs, which contributes to the operation of an organism. In fact, Ernst Mayr divided biology into two basic subdisciplines—functional biology and evolutionary biology. However, like the concept of purpose, function is not part of the vocabulary of physics. Atoms, electrons, molecules, have no function. Function, however, being a more pragmatic concept, is less metaphysically loaded than purpose, so the term does not arouse the same conceptual discomfort.

So, given the fact that function does not exist in the inanimate world (setting aside the functional entities that living things create to advance their agenda), how did function first arise during the emergence of life process? What was nature's primary function? As discussed in chapter 8, there is a causal connection between replicative function, complexity, and replicator stability, as was illustrated in figure 8.5. Thus, life's primary function can be identified as replication given its central role in the establishment of the replicative DKS state. Replication is the axis on which the entire life phenomenon revolves. Of course, biology is replete with functions other than replication—from ribosomes that churn out polypeptide chains through to pumping hearts. However, all such functions are secondary functions, functions that derive, directly or indirectly, from the primary function, the replicative function on which the entire life phenomenon rests, and from which they would have emerged.

To sum up: change in nature is invariably directed—toward persistent forms—and that directedness applies equally within the living world. Life's ultimate purpose, its goal, is to persist. The term *purpose* captures this drive in biological language; the term *persistence* captures it in physical language. To deny life's goal-directed character is to deny its essence. Moreover, it is the divergent nature of the paths to persistence in biology that makes the living world relatively unpredictable and fundamentally distinct from the "regular" chemical world. Indeed, it is precisely the generality and variability of the concept of purpose that makes it so useful. Life's fundamentally purposeful character thus offers deep insight into the nature of change in the living world, even if that directedness remains implicit and largely qualitative. The issue of purpose's subjective dimension—the problem of intentionality—is a separate question, a facet of the problem of mind, and the subject of the next chapter.

10
Physicalizing Mind

> Our life is shaped by our mind; we become what we think.
>
> —Gautama Buddha

Consciousness and mind are the most baffling of life's extraordinary properties. Consciousness, the ability to have subjective feelings, desires, intentions and complex thoughts, is so immediate, so intimate, yet in a physical sense so inexplicable. Mind, the more general term that refers to the entire system of thought, perception, memory, and problem solving, is no less confounding. How can matter—ultimately, just a bunch of atoms and molecules—think?

In his book *The Astonishing Hypothesis*, Francis Crick concluded that all aspects of mind and consciousness derive from physical processes in the brain. As he put it: "*you're nothing but a pack of neurons*," and in us humans, the "pack" can be spectacularly large, some eighty-six billion of them. But that large number already hints at how the problem of mind might be addressed. Brains did not start off with billions of neurons but evolved over time from simpler systems. The secret of mind may emerge through consideration of those earlier simpler systems.

Modern neuroscience has applied modern physical techniques to probe the nature of mind. Functional magnetic resonance imaging (fMRI) and electroencephalography (EEG) have been used to study activity in different regions of the brain (fMRI by measuring blood oxygen levels, EEG by measuring electrical activity) seeking to correlate those measurements with corresponding mental activity. Such neural correlates have taught us that neural activity is graded, that it's not just in a simple "on/off" state, that it is highly networked, and that emotion and rational thought are deeply intertwined. The fact that neural correlates have been able to identify regions of specific brain activity with corresponding mental activity supports Crick's view that mental activity ultimately has its origins within physical systems. But the mystery of how matter can give rise to feelings, intentions and complex thoughts remains just that—a deep physical mystery.

David Chalmers, the noted philosopher, raised the bar to understanding yet another notch through his formulation of the "hard problem of consciousness." Why is all that complex neural activity accompanied by conscious experience? Why does it feel like something to be you? Chalmers further argued that no matter how deeply we investigate the mechanistic workings of the brain, there is an unbridgeable gap between the objective nature of the physical processes that can be measured by fMRI or EEG, and subjective experiences, such as feeling pain, seeing red, or enjoying a concert. We know what it feels like to be ourselves, but, as Thomas Nagel, the New York University philosopher, pointed out, it presumably feels like something to be a bat, something which we cannot know.

A wide set of proposals to explain consciousness, ranging from mathematical to philosophical, even quantum mechanical, have been proposed over the years, and suffice it to say that the topic remains highly contentious. Jerry Fodor, an American philosopher of mind, once summed up the state of the field with his comment that "nobody has the slightest idea how anything material could be conscious," and Princeton neuroscientist Michael Graziano has argued that "most popular theories of consciousness are worse than wrong . . . they don't actually explain anything." With those dismissive background comments in mind, clearly one should proceed cautiously in the formulation of new ideas.

A popular view in the late twentieth century, that the brain could be considered analogous to a computer that processes information acquired from the external environment and produces output as thoughts, behaviors, and actions, did little to resolve the dilemma. That approach, which views the brain as a biological computer, fails to address the central question of how a subjective world of purpose, intentionality, desire, and emotion is at all possible. Certainly, computers do not give any indication they experience subjective feelings. The scientific fallback position continues to be that mind is fundamentally materially based, that life's mental capabilities somehow lie hidden within the physical domain. But how? Let me now discuss how the DKS concept may open a means of connecting between life's physical and mental facets.

Is Mental Reality Reducible at All?

Throughout this text, I have emphasized the power of reductionist methodology in seeking to understand complex systems. Recall our flight analogy: If you want to understand the principles of flight, study the Wright Brothers' 1903 airplane model, not a Boeing 747. Well, that same approach can usefully be applied to the study of mind. Rather than address the nature of mind in highly evolved life forms, let us trace the evolutionary process to its earlier stages when mind would have been more rudimentary, and therefore more readily characterizable. The strategy: *go simple.* Eric Kandel, the Columbia neuroscientist, adopted this "go simple" approach in his study of the mechanisms of memory and learning by investigating the nervous system of a simpler organism, the sea slug *Aplysia.* His work yielded surprisingly powerful insights into the neural mechanisms underlying short- and long-term memory. More fundamentally, however, if mind emerged from nonmind through an evolutionary process, its physical basis *must lie hidden within that very process.*

As a first step in following the "go simple" approach, let's switch our attention from the study of animals with their complex brains to the study of simple life, bacteria. Bacteria, it turns out, are full of surprises, not just in their physical behavior, but in their mental activity as well. In fact, recent microbiological research has turned much established neurological understanding on its head. Single cell organisms, lacking any neural structures of any kind, already possess strikingly sophisticated problem-solving and decision-making capabilities. Michael Levin terms this *nonneural intelligence* and there is growing support for that characterization. We have already noted James Broach's claim that microorganisms can "think," while James Shapiro has argued that bacteria are small but not "stupid." And Pamela Lyon described in marvelous detail the highly developed cognitive capabilities of bacteria.[1]

But let us remind ourselves why life has those mental capabilities—to survive, or as I prefer to put it, to persist. All life characteristics have emerged to advance that most basic physical endeavor.

The realization that bacteria are able to express complex mental activity means that the human brain's extraordinary complexity, with its eighty-six billion neurons and a similar number of nonneural cells, is not an essential requirement for persistent mental activity. Thus, just as bacterial life can excrete waste without kidneys, so bacteria can "think" without a brain. Simplest life with *no* neural structures of any kind already possesses strikingly sophisticated mental capabilities, suggesting that life's mental capabilities were initiated very early in the evolutionary process.

But if we believe in a continuous evolutionary process whose roots lie in chemistry, then life's fundamental characteristics, *including* its mental dimension, may well have started in chemistry. Is it possible that certain chemical systems already exhibit proto-cognitive characteristics out of which biological cognition could have evolved?[2]

The unexpected answer now appears to be *yes*. Life's mental capability, as all life capabilities, likely exists on a continuum, extending from minimal through to advanced, with its minimal level being rooted within a basic chemical structure. And, of course, once a basic chemical system with minimal mental capability was established within a system that was replicative, an evolutionary process would proceed to enhance that limited capability for improved functionality. Importantly, identifying the nature of that minimal level might offer physical insights into the origin and nature of mind. However, to extend the mental dimension to chemistry, we need a biological marker for that mental dimension, one that can be sought within chemistry. *Cognition*, one of life's fundamental characteristics, fits the bill. Let's unwrap what that basic concept entails.[3]

The Chemical Origin of Cognition

Cognition, like many concepts in biology, is not easily defined. Sara Shettleworth, a Canadian psychologist, however, has proposed: the mechanisms by which animals acquire, process, store, and act on information from the environment. These include perception, learning, memory, and decision-making.[4] It is a definition that has gained wide acceptance in part because though it was proposed for neural animals, it now appears equally applicable to all life forms, including nonneural microorganisms. That is quite remarkable. Just think, cognition in elephants and bacteria may well be defined in the same terms! This commonality regarding life's mental capability strengthens the growing view that cognition can be understood as fundamental to all life.

At the heart of the Shettleworth definition lies the concept of *information*. Information itself is an elusive concept, a common source of scientific confusion, as its definition depends on the scientific discipline that seeks to employ it. As part of information theory, the concept begins with Claude Shannon's landmark idea that information is the "reduction of uncertainty" in a system, and that it can be quantified and transmitted without regard to the nature of the information that is provided. Entropy is the measure of the uncertainty or unpredictability in a system—the greater its uncertainty, the greater its entropy.[5] Let's consider a simple example.

Without describing the math by which informational entropy is measured, throwing a dice provides 2.585 bits of information, while tossing a coin provides just 1.0 bits of information—a dice throw is more informative than a coin toss. Why the difference? There are more possible outcomes for a dice throw (six possibilities) than for a coin toss (two possibilities). Accordingly, a dice throw reduces uncertainty to a greater extent than a coin toss. To repeat, Shannon's theory does not deal with the nature of information, just its quantity, its ability to surprise. Interestingly, through Shannon's ideas on information, the concept of entropy has evolved from being just a physical/thermodynamic one (as proposed by Clausius and Boltzmann in the nineteenth century), into a mathematical/probabilistic one.

In biology, the concept of information is of a more practical kind—it is the information itself that's important, rather than its quantity. Information in biology has been defined as whatever induces a change in a living thing's physiology or its interactive potential with its environment. But our presumption that biology emerged from chemistry in a continuous evolutionary process suggests that early in the evolutionary process, some chemical entity would have been able to receive and respond to information in a biologically relevant way.

Well, here's the surprise. Based on the Shettleworth definition, chemical DKS systems, such as van Esch's dynamic hydrogels, already express the beginnings of what could be considered proto-cognitive capabilities.

Such systems interact and exchange information with their environment, just like biological systems. By definition, therefore, one might say they express in some minimal sense "awareness" of their environment. Furthermore, being in that DKS state, they adapt to perturbations that might be applied. Thus, just like biological entities, chemical DKS systems respond to changing circumstances. And let's not forget that chemical DKS systems do the chemical equivalent of "eat," "excrete waste" and "die." As already noted, in what is clearly just a chemical system, we can already discern central characteristics associated with "aliveness."

But here's the thing: the cognitive relationship between any DKS entity, chemical or biological, and its environment is not just physical but begins to express mental aspects as well. Let us remind ourselves what "mental" means. In the most basic terms, mental activity can be defined as information-processing activity—the receiving, transforming, storing, and utilizing of information within a self-organizing system. Thus, a chemical DKS system could also be considered as expressing the beginnings of rudimentary mental activity—one might term it proto-mental activity—and in what is unambiguously a physical/chemical system. Having said that, a DKS chemical system, such as the van Esch hydrogel, should not be designated as being mentally active. Just as we wouldn't characterize a leaf caught up in a gust of wind as flying, we shouldn't characterize a chemical DKS system as being mentally active.

But once the DKS system becomes replicative, we are in a new ballpark. The existence of matter that is both replicative *and* in the DKS state is the game changer. Replication would induce an evolutionary process toward more persistent replicators and that rudimentary mental potential would become magnified for its functional advantages. Classic Darwinian natural selection would begin to operate. Thus, one would expect that explicit mental characteristics would begin to emerge from what was previously "just" a physical system. Crucially, rudimentary mental activity would have been present at the outset of the evolutionary process, just as Darwin envisaged well over a century ago. The physical/chemical basis of life's mental dimension may therefore be identified. *It appears to lie in the dynamic interactive nature of the DKS state.*

This way of thinking fits in comfortably with the biological theory of autopoiesis, formulated in the 1970s by Humberto Maturana and Francisco Varela, two pioneering Chilean biologists.[6] Their theory characterized life as an autonomous dynamic system that maintains itself by creating its own components and identity through continual interaction with its environment. Their biological perspective reaches similar conclusions to those obtained from our physical/chemical view. Maturana's insightful comment that "living as a process is a process of cognition" hits the nail on the head. Both the theory of autopoiesis and life's DKS characterization affirm that cognition is intrinsic to life, but the DKS description goes a step further by revealing how the process has its roots in chemistry.

Notice, also, that a chemical DKS system, though ontologically distinct from its environment, is not structurally constrained by its environment. In fact, the reverse is true—it is the dynamic and ongoing interaction of the DKS system with its environment that enables the evolutionary process to progress, leading the DKS system to continually adapt to its environment, seeking ever more persistent forms. Any structural modification that would serve to enhance the system's stability within its environment would be facilitated in accord with the persistence principle. Thus, the emergence of life's cognitive capabilities within chemical replicative DKS aggregates would have accompanied life's Big Bang moment—the emergence of a chemical replicative DKS system—the moment from which complexification toward explicit life forms with enhanced cognitive capabilities would have taken place.

The above discussion outlining the physical basis of mind may well have profound implications. Just as the chemical discovery of kinetic states of matter opened up new horizons in chemical possibility, the existence of such chemical states opens the door to the likelihood of previously unrecognized physical states—*physical states of mind.* At the 13th Nobel Conference, held in 1977 in St. Peter, Minnesota, Max Delbrück, the Nobel biophysicist, spoke of the "feeling of absurdity that attaches to the notion 'mind from matter.'" Yet, despite the enormous strides taken by physics in the twentieth century with the discovery of theories of relativity and quantum mechanics, the physics of mind has remained an impenetrable mystery. Through the reality of kinetic states of matter, physics may finally begin to extend its reach deeper into biology by beginning to address its most mysterious aspect—the physics of mind.

Origin of Consciousness

Few would dispute that, before life appeared on Earth, consciousness simply was not there. At some point in deep time, mind arose from matter. That claim alone forces us to confront one of science's deepest puzzles: How could a universe, seemingly mindless at its foundation, give rise to entities that experience, feel, choose, and know? Setting aside panpsychism—the idea that consciousness pervades the universe in rudimentary form—we are confronted with three hard questions: What is consciousness? How did it emerge? And why? Let us build on the DKS framework to see how these questions might be addressed. First, what is it?

Barbara McClintock, in her groundbreaking work on cellular behavior almost a century ago, offered more than a clue. She observed that cells do far more than carry out preset biochemical routines. They monitor their internal condition, sense changes, evaluate damage, and restructure themselves as needed. In her language, cells possess a kind of "sense of self": an intrinsic awareness when something inside has gone awry and needs to be restored.[7] Cells, in short, *know*—in their own limited but real way—"what's out there" and "what's in here." Also, with time, it has become increasingly clear that her descriptions were not intended purely metaphorically (as discussed in the epilogue). If so, it is then not a great leap to think of such "self-knowledge" as a rudimentary form of consciousness: a primordial, utilitarian awareness that emerged because it helped living systems persist. But if this is true, what is its physical anchor?

We previously described how a chemical DKS system, such as van Esch's dynamic hydrogel, expresses primitive proto-cognitive capabilities through dynamic interaction with its supportive environment. But that suggests that a rudimentary form of self-awareness already exists within a *chemical* DKS system. As noted, all DKS systems "know" in a limited sense "what's out there" and "what's in here." In fact, it is only on that basis that DKS systems are able to maintain their dynamic persistent state.

Of course, once such a system becomes *replicative*, natural selection would tend to amplify these capabilities. Evolution's relentless filtering pushes DKS systems toward ever-increasing persistent forms—forms that monitor themselves *more* effectively, respond *more* flexibly, and ultimately make *more* sophisticated "decisions." The trajectory points toward increasing self-awareness, step by incremental step. From this perspective, consciousness is not an inexplicable add-on to life but an outgrowth of life's physical strategy for persisting in a kinetic regime.

But what about the "feel" of consciousness—the redness of red, the sting of pain? That dimension is necessarily private. Just as one cannot penetrate the subjective experience of another person, the inner experience (if any) of a bacterium is forever closed to us. We infer its "feelings," if we can call them that, only through behavior. In any case, the level of consciousness we humans experience with our eighty-six billion neurons should not serve as a guide in attempting to assess how a bacterium might "feel" based on a far simpler DKS system lacking neuronal structures of any kind. The criterion for operative self-awareness or consciousness should not be based on inaccessible subjective feel, but rather on behavioral and functional responses to cognitive input.

The above comments should not be interpreted to mean that a chemical DKS system, such as the van Esch hydrogel, experiences limited consciousness. Such chemical DKS systems are described to help identify the chemical basis and origin of "self." It is only once the level of self-awareness evolves to a level approaching that found in simplest (bacterial) life that it facilitates decision-making and problem-solving capabilities. Seen this way, consciousness lies on a continuum, absent in rocks or crystals, vanishingly faint in microbes, and elaborated massively in creatures with complex nervous systems.

The underlying message should be clear: it is the physical existence of DKS systems that enables the dimension of "self-awareness" to emerge, one that underpins the phenomenon of consciousness. Of course, the knowledge of precisely where on that axis of self-awareness any cognitive entity might lie will likely remain out of reach. One cannot hope to quantify a phenomenon that is inherently both inaccessible and unquantifiable.

But once it is recognized that a basic level of consciousness is intrinsic to the life phenomenon, it should not surprise us that consciousness at some level would likely extend throughout the living kingdom. We humans frequently like to view ourselves as distinct to other species, but that exceptionalism is now widely doubted. In *Beyond Words*, Carl Safina describes how even the range of human emotions—jealousy, anger, love, grief, joy—are evident in a range of animal species, elephants, wolves, even killer whales. We should not be surprised. Evolution, after all, is a master of adaptation, continually reusing, reshaping, and fine-tuning existing capabilities, which often result in convergent evolutionary outcomes across species—recall the multiple kinds of eye.

The message is clear: a replicative DKS system provides a physical foothold for life's mental dimension together with its subjective expression, consciousness. It offers a route by which self-awareness—not as mysticism, but as chemistry directed by evolution—could emerge. As philosopher Massimo Pigliucci has put it, consciousness is neither a "spooky mystery" nor mere illusion; it is an established biological phenomenon with real causal power.[8] It evolved because organisms that could monitor themselves, interpret their surroundings, and choose wisely were more likely to survive. Cognition and consciousness are thus two sides of the same evolutionary coin. Cognition marks the objective, information-processing side; consciousness signals the subjective, internal vantage point. Both grow out of the replicative DKS state—life's kinetic response to a universe inexorably pulled toward equilibrium and thermodynamic finality. Indeed, if this DKS view of the mind is correct, the nature of life's nonphysical aspects may well be a step closer to physical resolution. As Francis Crick predicted several decades ago, mind may not be an enigma beyond nature but a coherent, lawful phenomenon rooted in chemistry, shaped by evolution, and powered by the enduring struggle of living systems to persist in time.

The Nature of Intelligence

We have already noted that intelligence—broadly defined as the capacity of a system to generate adaptive behavior in response to its environment—is already manifest in the simplest of life forms. But given recent advances in the study of artificial intelligence (AI), we might consider how the two possible kinds of intelligence that seemingly exist—artificial and biological—relate to one another.

Biological intelligence (BI) derives from the fundamental reality that the living cell continuously regulates itself to stay alive through dynamic chemistry. The cell continually repairs, rebuilds, and responds to its surroundings, thereby expressing intelligent behavior—not as conscious thought in the human sense, but through embodied and self-maintaining regulation. BI is of a dynamic kind, operating from within the system and driven by the need to survive.

By contrast, artificial intelligence (AI), as implemented in large language models (LLMs), is very different—it is built from externally imposed architectures. LLMs do not metabolize, self-repair, or reproduce, and have no intrinsic stake in their own persistence. Instead, they learn by absorbing vast quantities of human-generated information and extracting statistical regularities from within that information. Thus, whereas biological systems generate functional knowledge in the service of continued existence, computers generate patterns that we interpret as meaningful without the computational system itself possessing any intrinsic awareness of what those patterns signify. The two intelligence types therefore differ fundamentally in their grounding: BI is rooted in the logic of life—dynamic, embodied, self-maintaining, self-aware—whereas AI is rooted in mathematical abstraction and algorithmic computation. The brain is not a computer made of meat.

Which leads us to the dilemma of free will. Does it exist? Are cognitive animals—humans in particular—

free to choose between possible courses of action, or are all such decisions determined by prior causes beyond the living entity's control? The divergent topology of replicative DKS space (figure 8.4) offers a possible means of addressing this question. Any replicative system in the DKS state, a state that is inherently autonomous, can potentially choose between different paths towards increasing DKS due to the topology's divergent nature. Thus, from a DKS perspective, what we call "free will" may be reconceptualized as constrained choice within a divergent space of possible stability-enhancing paths—autonomy expressed as selection among real alternatives. These ideas need to be explored further, but here also, biology's inclination to explain away unexplained characteristics as "illusory" could be questioned.

Finally, Barbara McClintock's insights into cellular intelligence and awareness are helpful in investigating the relative importance of physical and mental capabilities along the evolutionary path. Her early thinking provides a useful framework for understanding how early mental capabilities might have offered greater marginal benefits over physical adaptations, enabling proto-consciousness to emerge and evolve. Looking back, it is evident that it was an evolutionary process along a coupled physical/mental axis that led to the emergence of unprecedented human technological capabilities, enabling life forms (us) to uncover nature's laws, to ponder on the origin of life and the earliest moments of the physical universe. The human species' control of this planet did not derive from human life's physical attributes, but from its mental ones. Natural selection's most profound "discovery" was that of the mental domain. Once that domain was established, an evolutionary process along that mental axis became the overriding imperative. Indeed, the human brain, weighing just two percent of total body weight in an average adult, consumes approximately twenty percent of the body's total energy consumption, pointing to the relative importance of mental function.

Surveying the current state of our planet makes it clear that life's mental capabilities were far more influential in shaping the planet's geophysical and biological character than its purely physical ones. An expression attributed to Buddha captures that idea well: "Our life is shaped by our mind; we become what we think." That expression now looks to extend beyond the metaphorical. In fact, the future of our planet—and the living world that it supports—may ultimately depend on how the evolutionary process comes to direct life's mental capacities. The social and evolutionary significance of our entering a new phase in which biological and artificial intelligence can combine to advance life's broader aims is difficult to assess. As emphasized earlier, it is impossible to predict where a divergent trajectory will lead. Nonetheless, the coupling of these two forms of intelligence suggests that we are on the cusp of a "brave new world." Whether we humans will remain in control of that world is a daunting, but separate question.

Epilogue

I have attempted to take you on a chemical journey to uncover the mystery of life. Just as we understand why water is liquid, metals are shiny, and salt is crystalline, we can now begin to understand why life exists, why it is purposeful, self-aware, mindful, and complex.

So what held us back all this time? Why has the life problem proven so intractable?

Paradoxically, biology's commitment to a strict physicalist view of life looks to have been the major impediment to progress. The seventeenth century "nature is objective" worldview that revolutionized scientific understanding was so intellectually compelling, so conceptually intoxicating, that it induced the natural sciences, and biology in particular, to marginalize any biological fact or explanation that did not fit in with that grand physical view. The reality of life's nonphysical characteristics was dismissed as either "illusory" or brushed aside as "emergent properties" or simply set aside for another day. No deviation from the physicalist worldview was deemed acceptable.

But with the benefit of hindsight, the view that "nature is objective" appears to have been overstated. The clockwork metaphor—treating the universe as a predetermined, machine-like system—failed to capture reality in its entirety. With the phenomenon of life, subjectivity is not ambiguous or incidental; it is unmistakable. Yet it has long been deemed acceptable to assign life a separate, even contrarian, set of rules—without seriously reexamining the axiom of objectivity itself. Physical laws, however, cannot be applied selectively. When their application produces internal inconsistencies, those inconsistencies point not to deviations within natural law but to shortcomings in the laws as formulated. Any coherent attempt to uncover nature's global principles must therefore begin from a more fundamental premise: nature is one.

Looking back a century, there was one voice that boldly chose to break out of the physicalist straitjacket that biology had imposed on itself. I refer to that iconic researcher, Barbara McClintock, mentioned frequently in this text. Her commitment to accept empirical reality, no matter how puzzling and unexpected, led her to question central biological assumptions from the earliest days of the modern synthesis. She questioned the gene-centered view of life that considered the genome to be a static entity, supposedly serving as a molecular instruction manual for the organism. Her work on transposable elements (so-called "jumping genes") in the late 1940s and early 1950s led her to the realization that the cell's genome is dynamic, interactive, and able to reorganize itself. Indeed, in a giant conceptual leap, she reached the profound insight that it was not the genome that controlled the cell, but rather the cell that controlled its genome. Though her conclusions were derived from years of experimentation and observation, the views she expressed conflicted with established biological thinking, causing her work to be ignored, even discredited. It took several decades before her studies on mobile DNA elements were duplicated in several additional life forms, thereby validating her earlier work and earning her the 1983 Nobel Prize in Medicine or Physiology.

Even then in her Nobel Prize lecture, she took the argument a bold step further. On the Nobel podium she declared: "A goal for the future would be to determine the extent of knowledge the cell has of itself, and how it utilizes this knowledge in a 'thoughtful' manner . . ." Self-knowledge? Utilizing knowledge thoughtfully? Some in the audience would have felt quite uncomfortable with such nonphysicalist thinking.

And yet, in a clear-eyed acknowledgment of the cell's purposeful character, she was pointing to where the true mystery lay. The cell was not a passive molecular machine, but an active participant in its own fate—able to recognize danger, initiate appropriate responses, and structurally adapt to maintain itself. The real mystery lay in how such a capability could have emerged from inanimate beginnings. On this point, however, neither physical nor biological answers were forthcoming.

Fast forward to recent times. I believe the discovery of the existence of *kinetic states of matter* is beginning to impact on our understanding of the life phenomenon no less dramatically than the discovery of molecules and thermodynamic states had on our understanding of traditional chemistry over a century ago. In this new light, autonomy, self-awareness, adaptiveness, purposefulness, and mind are finally beginning to make physical/chemical sense. Yes, much of nature is objective and governed by natural law, as recognized several centuries ago by Newton, Galileo, Descartes, and company. But there are regions within chemical space that are *subjective, purposeful,* and *mindful.* Repurposing Richard Dawkins's "selfish gene" metaphor, one could think of the kinetic state of matter as the *selfish state,* the physical state of matter that opens the door to subjective matter and subjective behavior. And once subjectivity evolved and intensified, elements of the universe began to study and explore other elements of the universe—the emergence of kinetic states of matter enabled the universe to discover itself!

So, given these new insights, can one make some predictions regarding the evolutionary future? As noted, recent advances in artificial intelligence (AI) suggest that we are entering an explosive era of technological capability, perhaps far more profound than the one we have recently experienced. Once AI begins to direct its own development, we will be on an exponentially driven technological track, necessarily unsustainable.

Human nature, fundamentally subjective and molded over millennia to deal with life challenges of the classical kind, will struggle to adapt to such dramatically changing life dynamics.

These thoughts bring to mind the Fermi paradox, as posed by the physicist Enrico Fermi in 1950. During a lunchtime conversation at the Los Alamos National Laboratory, he famously asked "where is everybody?" Given our view of life as a persistent chemical form maintained by the kinetic power of exponential growth, and given the immensity of the cosmos, it seems statistically inconceivable that life in one chemical form or other would not have emerged in other locations in the universe. But persistence and high intelligence may well be incompatible. Intelligence may confer local power while undermining long-term stability. Could it be that human society, bestowed with runaway machine intelligence, could become kinetically unstable? Could this fate be common in the universe? We cannot say for sure. Alien or terrestrial, one cannot predict where a divergent path will lead.

Regardless, at a personal level, I write this book with a feeling of considerable privilege that some four score years ago, a bunch of molecules dynamically aggregated to become me, and managed to persist, through thick and thin, thereby providing me with a window seat on a voyage of discovery on this latest and most dramatic segment of life's four-billion-year odyssey. I am especially indebted to the companionship of family and friends that accompanied me on this dramatic, occasionally bumpy, journey. Ultimately, any journey is only as memorable and worthwhile as the company with which it is shared, and in that regard, I have been profoundly fortunate. Looking ahead, I contemplate with some trepidation the challenges we humans collectively face as we enter the most unpredictable, and, possibly, the most uncertain phase of our 300,000-year sojourn on this earthly lump of matter we call home. In seeking persistence, life presses on unremittingly, relentlessly. Fellow travelers, *bon voyage*!

Acknowledgments

Many of the ideas described in this book have emerged based on discussions, interactions, and collaborations with scientists from a range of disciplines. Life's origins and nature are as much about physics, chemistry, information theory, and computer science as they are about biology. First and foremost, I am indebted to my longstanding colleague Robert Pascal, Aix-Marseille University, whose deep understanding across the sciences enabled a wonderfully productive collaboration over many years. His critical thinking was invaluable in the formulation of concepts able to link the physical and biological sciences. As a chemist, my many discussions with biologists over the years have been invaluable in the formulation of the ideas described here. In particular, I want to thank Peter Corning, Tim Coulson, David Deamer, Frank Harold, Kevin Lala, Puri López-García, Bill Miller, James Shapiro, John Odling-Smee, Scott Turner, and Dick Vane-Wright, all of whom helped me contend with the physics-biology conceptual divide. I am indebted to my brother-in-law Alon Orlitsky, an information theorist who explained how "surprise" can be quantified, and also to Steve Grand and Blaise Agüera y Arcas, who separately showed me how computational logic can provide important physical insights. Discussions with Sean Carroll, Sijbren Otto, Leo Radom, Andrew Sundstrom and Eörs Szathmáry, and conceptual insights from David Chalmers, Stuart Kauffman and John Mattick are also gratefully acknowledged. I thank my sister, Elinor O'Neill, her husband, David, for a careful reading of early drafts of the manuscript and for pointing out inconsistencies and obscurities, and my son Guy for his assistance in producing the figures. Last, but not least, I am indebted to my wife, Nella, whose encouragement and support were invaluable. Her critical faculties kept many of the central arguments coherent and precise. The two children we raised and the five grandchildren we were gifted with were a constant reminder what the life phenomenon is about, and what life's purpose, supposedly a scientifically contentious issue, is at its core. Finally, I'm indebted to the team at the MIT Press, in particular to Stephen Morrow for his many thoughtful suggestions and critical editorial support and to Caroline Helms for her technical assistance.

Notes

Chapter 1

1. Sean M. Carroll, *The Big Picture* (Dutton, 2017).

Chapter 2

1. James Watson and Francis Crick, "Genetical Implications of the Structure of Deoxyribonucleic Acid," *Nature* 171 (1953): 964–967.

2. Richard Dawkins, *The Selfish Gene* (Oxford University Press, 1989).

3. Kevin Laland, Tobias Uller, Marc Feldman, Kim Sterelny, Gerd B. Müller, Amin Moczek, Eva Jablonka, John Odling-Smee, "Does Evolutionary Theory Need a Rethink?" *Nature* 514 (2014): 161–164.

4. Ernst Mayr, *Toward a New Philosophy of Biology* (Harvard University Press, 1988).

5. Ernst Mayr, "Teleological and Teleonomic, A New Analysis," in *A Portrait of Twenty-Five Years: Boston Studies in the Philosophy of Science*, ed. Robert S. Cohen and Marx W. Wartofsky (Springer, Dordrecht, 1974).

6. Richard Dawkins, *The Blind Watchmaker: Why the Evidence of Evolution Reveals a Universe without Design* (Norton, 1996).

7. Eric R. Kandel and James H. Schwartz, "Molecular Biology of Learning: Modulation of Transmitter Release," *Science* 218 (1982): 433–443.

8. Jacques Monod, *Chance and Necessity* (Random, 1971).

9. Aaron Ciechanover, Hannah Heller, Sarah Elias, Arthur L. Haas, and Avram Hershko, "ATP-Dependent Conjugation of Reticulocyte Proteins with the Polypeptide Required for Protein Degradation," *Proceedings of the Natural Academies of Sciences of the United States of America* 77 (1980): 1365–1368.

10. John Mattick and Paulo P. Amaral, *RNA, the Epicenter of Genetic Information* (Taylor & Francis, 2022).

11. Barbara McClintock, *Discovery and Characterization of Transposable Elements: The Collected Papers of Barbara McClintock* (Garland, 1987).

12. James A. Shapiro, "The Basic Concept of the Read-Write Genome: Mini-Review on Cell-Mediated DNA Modification," *Biosystems* 140 (2016): 35–37.

13. Denis Noble, "The Illusions of the Modern Synthesis," *Biosemiotics* 14 (2021): 5–24.

14. Santosh Manicka and Michael Levin, "Modeling Somatic Computation with Non-Neural Bioelectric Networks," *Scientific Reports* 9, 18612 (2019).

15. Simon Conway Morris, "Evolutionary Convergence," *Current Biology* 16, R826 (2006).

16. Carl R. Woese, "A New Biology for a New Century," *Microbiology and Molecular Biology Review* 68, no. 2 (2004): 173–186.

Chapter 3

1. Carl R. Woese and George E. Fox, "Phylogenetic Structure of the Prokaryotic Domain: The Primary Kingdoms," *Proceedings of the National Academies of Sciences of the United States of America* 74, no. 11 (1977): 5088–5090.

2. Stanley L. Miller, "A Production of Amino Acids Under Possible Primitive Earth Conditions," *Science* 117, no. 3046 (1953): 528–529.

3. Bhavesh Patel, Claudia Percivalle, Dougal J. Ritson, Colm D. Duffy, and John D. Sutherland, "Common Origins of RNA, Protein and Lipid Precursors in a Cyanosulfidic Protometabolism," *Nature Chemistry* 7 (2015): 301–307.

4. J. Baz Jackson, "Natural pH Gradients in Hydrothermal Alkali Vents Were Unlikely to Have Played a Role in the Origin of Life," *Journal of Molecular Evolution* 83 (2016): 1–11.

5. David W. Deamer, *Assembling Life*: How Can Life Begin on Earth and Other Habitable Planets? (Oxford University Press, 2019).

6. Tibor Ganti, "Biogenesis Itself," *Journal of Theoretical Biology* 187, no. 4 (1997): 583–593.

7. D. R. Mills, R. L. Peterson, and S. Spiegelman, "An Extracellular Darwinian Experiment with a Self-Duplicating Nucleic Acid Molecule," *Proceedings of the National Academies of Sciences of the United States of America* 58, no. 1 (1967): 217–224.

8. Leslie E. Orgel, "Evolution of the Genetic Apparatus," *Journal of Molecular Biology* 38, no. 3 (1968): 381–393; Francis Harry Compton Crick, "The Origin of the Genetic Code," *Journal of Molecular Biology* 38, no. 3 (1968): 367–379; Carl R. Woese, *The Genetic Code*: The Molecular Basis for Genetic Expression (Harper and Row, 1967): 179–195.

9. Thomas R. Cech, "The Ribosome is a Ribozyme," *Science* 289, no. 5481 (2000): 878–879.

10. Ada Yonath, "High-Resolution Structures of Large Ribosomal Subunits from Mesophilic Eubacteria and Halophilic Archaea at Various Functional States," *Current Protein & Peptide Science* 3, no. 1 (2002): 67–78.

11. Stuart Kauffman, *At Home in the Universe*: The Search for the Laws of Self-Organization and Complexity (Oxford University Press, 1995).

Chapter 4

1. Claude Shannon, "A Mathematical Theory of Communication," *Bell System Technical Journal* 27 (1948): 379–423.

2. Martin Jinek, Krzysztof Chylinski, Ines Fonfara, Michael Hauer, Jennifer A. Doudna, and Emmanuelle Charpentier, "A Programmable Dual-RNA-Guided DNA Endonuclease in Adaptive Bacterial Immunity," *Science* 337, no. 6096 (2012): 816–821.

Chapter 5

1. Murray Gell-Mann, "What is Complexity," *Complexity* 1, no.1 (1995).

2. Erwin Schrödinger, *What Is Life?: The Physical Aspect of the Living Cell* (Cambridge University Press, 1944).

3. Steve Atkinson and Paul Williams, "Quorum Sensing and Social Networking in the Microbial World," *Journal of the Royal Society Interface* 6, no. 40 (2009): 959–978.

4. Stephen Jay Gould, *Full House: The Spread of Excellence from Plato to Darwin* (Harmony, 1996).

Chapter 7

1. Alfred J. Lotka, "Natural Selection as a Physical Principle," *Proceedings of the National Academies of Sciences of the United States of America* 8, no. 6 (1922): 151–154.

2. Steve Grand, *Creation: Life and How to Make It* (Harvard University Press, 2001).

3. Addy Pross and Vladimir Khodorkovsky, "Extending the Concept of Kinetic Stability: Toward a Paradigm for Life," *Journal of Physical Organic Chemistry* 17, no. 4 (2004): 312–316; Addy Pross, "Stability in Chemistry and Biology: Life as a Kinetic State of Matter," *Pure and Applied Chemistry* 77, no. 11 (2005): 1905–1921.

4. Job Boekhoven, Wouter E. Hendriksen, Ger J.M. Koper, Rienk Eelkema, Jan H. van Esch, "Transient Assembly of Active Materials Fueled by a Chemical Reaction," *Science* 349, no. 6252 (2015): 1075–1079.

5. Andreas Walther, "Viewpoint: From Responsive to Adaptive and Interactive Materials and Materials Systems: A Roadmap," *Advanced Materials* 32, no. 20 (2020).

Chapter 8

1. Job Boekhoven, Wouter E. Hendriksen, Ger J.M. Koper, Rienk Eelkema, Jan H. van Esch, "Transient Assembly of Active Materials Fueled by a Chemical Reaction," *Science* 349, no. 6252 (2015): 1075–1079.

2. In this text, we use the term *dynamic kinetically stable (DKS) state* to describe dynamic non-equilibrium structures such as the hydrogel product, while the literature often employs the term *dissipative self-assembly*. Though the latter term relates the phenomenon to Prigogine's term of *dissipative structures* (hurricanes, whirlpools, and so on), it marginalizes its relevance to the life phenomenon. The former term gives a clearer indication of its dynamic character, its relationship to traditional thermodynamic states, and its possible physical connection to the life phenomenon.

3. Robert Pascal and Addy Pross, "Chemistry's Kinetic Dimension and the Physical Basis for Life," *Journal of Systems Chemistry* 7, no.1 (2019): 1–8.

4. The dynamic balance between replicator formation and decay that forms the basis for the replicative DKS state can be represented by a simple differential equation:

$$dX/dt = kMX - gX$$

where X is the concentration of the replicator, M is the concentration of building blocks of which X is composed (for simplicity assumed to be of one kind), and k and g are rate constants for replicator formation and decay, respectively. Thus, the kMX term represents the rate of replicator formation from its building blocks, M, while gX represents the replicator's rate of decay. A steady state is established when $dX/dt = 0$.

5. Blaise Agüera y Arcas, Jyrki Alakuijala, James Evans, Ben Laurie, Alexander Mordvintsev, Eyvind Niklasson, Ettore Randazzo, and Luca Versari, "Computational Life: How Well-Formed, Self-Replicating Programs Emerge from Simple Interaction," https://doi.org/10.48550/arXiv.2406.19108.

6. Avram Hershko and Aaron Ciechanover, "The Ubiquitin System," *Annual Review of Biochemistry* 67 (1998): 425–479.

7. Shneior Lifson, "On the Crucial Stages in the Origin of Animate Matter," *Journal of Molecular Evolution* 44 (1997): 1–8.

Lifson described mathematically the result of *two* replicators, X_1 and X_2, in the DKS state competing for the same building blocks, M. The relevant kinetic equations now become:

$$dX_1/dt = k_1 M X_1 - g_1 X_1$$

$$dX_2/dt = k_2 M X_2 - g_2 X_2$$

The competitive steady state situation, obtained by simultaneously solving for $dX_1/dt = 0$ and $dX_2/dt = 0$, shows that the two replicators cannot coexist. The concentration of one of the replicators, either X_1 or X_2, drops to zero. The more DKS stable replicator is predicted to drive the less stable one into extinction.

8. Bin Liu, Juntian Wu, Marc Geerts, Omer Markovitch, Charalampos G. Pappas, Kai Liu, and Sijbren Otto, "Out-of-Equilibrium Self-Replication Allows Selection for Dynamic Kinetic Stability in a System of Competing Replicators," *Angewandte Chemie* 61, no. 18 (2022).

9. Richard Dawkins, *The Blind Watchmaker: Why the Evidence of Evolution Reveals a Universe Without Design* (Norton, 1996).

10. Simon Conway Morris, *Life's Solution: Inevitable Humans in a Lonely Universe* (Cambridge University Press, 2003).

11. Addy Pross, *What Is Life?: How Chemistry Becomes Biology* (Oxford University Press, 2016); Addy Pross, "Toward a General Theory of Evolution: Extending Darwinian Theory to Inanimate Matter," *Journal of Systems Chemistry* 2, no. 1 (2011): 1–14.

12. Stuart A. Kauffman, *Investigations* (Oxford University Press, 2000).

13. James Lovelock, *Gaia: A New Look at Life on Earth* (Oxford University Press, 1979).

14. W. Ford Doolittle, *Darwinizing Gaia* (MIT Press, 2024).

15. Peter A. Corning, *The Synergism Hypothesis: A Theory of Progressive Evolution* (McGraw-Hill, 1983).

Chapter 9

1. J. Scott Turner, *Purpose and Desire* (HarperOne, 2017).

2. Fallon Durant, Junji Morokuma, Christopher Fields, Katherine Williams, Dany Spencer Adams, and Michael Levin, "Long-Term, Stably Inherited Two-Headed Planaria Result from Modulating Bioelectric Networks," *Biophysical Journal* 112, no. 10 (2017): 2231–2243.

3. Sharad Ramanathan and James R. Broach, "Do Cells Think?" *Cellular and Molecular Life Sciences* 64 (2007): 1801–1804; Pamela Lyon, "The Cognitive Cell: Bacterial Behavior Reconsidered," *Frontiers in Microbiology* 6 (2015).

4. *Evolution "On Purpose": Teleonomy in Living Systems*, ed. Peter A. Corning, Stuart A. Kauffman, Denis Noble, James A. Shapiro, Richard I. Vane-Wright, and Addy Pross (MIT Press, 2023).

5. Martin Jinek, Krzysztof Chylinski, Ines Fonfara, Michael Hauer, Jennifer A. Doudna, and Emmanuelle Charpentier, "A Programmable Dual-RNA-Guided DNA Endonuclease in Adaptive Bacterial Immunity," *Science* 337, no. 6096 (2012): 816–821.

6. Erwin Schrödinger, *What is Life?: The Physical Aspect of the Living Cell* (Cambridge University Press, 1944).

7. Niels Bohr, "Light and Life," *Nature* 131 (1933): 457–459.

8. Eugene P. Wigner and Andrew Szanton, *The Recollections of Eugene P. Wigner* (Springer, 1992).

Chapter 10

1. Pamela Lyon, "The Cognitive Cell: Bacterial Behavior Reconsidered," *Frontiers in Microbiology* 6 (2015).

2. As an aside, Darwin, in his genius, already sensed that life's mental dimension began early in the evolutionary process. In Darwin's words, evolution took place through both "corporeal and mental endowments." Darwin's insight was simple but profound: mental activity was unlikely to have appeared out of the blue at some point along the evolutionary process. It had to be there from the beginning. Evolution is more about *improving* than *inventing*. Thus, the evolutionary process, from its outset, would have taken place along both physical *and* mental axes. One hundred and sixty years on, microbiological research is confirming Darwin's early intuition.

3. Robert Pascal and Addy Pross, "Toward the Physicalization of Biology: Seeking the Chemical Origin of Cognition," in *Evolution "On Purpose": Teleonomy in Living Systems*, ed. Peter A. Corning, Stuart A. Kauffman, Denis Noble, James A. Shapiro, Richard I. Vane-Wright, and Addy Pross (MIT Press, 2023).

4. Sara J. Shettleworth, *Cognition, Evolution, and Behavior* (Oxford University Press, 1998).

5. Claude Shannon, "A Mathematical Theory of Communication," *Bell System Technical Journal* 27 (1948): 379–423 (1948).

6. Humberto R. Maturana and Francisco J. Varela, *Autopoiesis and Cognition: The Realization of the Living* (Springer Dordrecht, 1980).

7. Barbara McClintock, *Discovery and Characterization of Transposable Elements: The Collected Papers of Barbara McClintock* (Routledge, 1987).

8. Massimo Pugliucci, "Consciousness is Real," *Aeon*, December 16, 2019, https://aeon.co/essays/consciousness-is-neither-a-spooky-mystery-nor-an-illusory-belief.

Index

Note: Page numbers in italics indicate references to figures.